Pryse Pryse (Baronet,

Gogerddan

Cardiganshire
S. Wales

1894

# HISTORY

OF

# THE ROYAL

# BERKSHIRE MILITIA

(NOW 3RD BATTALION ROYAL BERKS REGIMENT)

**Reading:**

PRINTED FOR THE AUTHORESS BY JOSEPH HAWKES.

COPIES MAY BE OBTAINED FROM MISS THOYTS, SULHAMSTEAD
PARK, BERKSHIRE, ALSO THROUGH BOOKSELLERS.

1897

# PREFACE.

——

YOU who are kind enough to read these pages, be lenient, I pray you, to the many errors doubtless contained therein; remember I am an antiquary, not a soldier, and the first woman who has ventured to write a Regimental History!

The search for information has been difficult; none was forthcoming from the regiment itself, as the only records they possess consist of a Register of Officers and two Court-martial Books, dating from 1803-1815, and two others of 1855-1861. An application to the Clerk of the Lieutenancy of Berkshire revealed nothing; an enquiry at the Bodleian Library gave the same answer. I wasted an hour at the British Museum, waiting vainly for books, or an answer as to whether they were in the Library.

The Record Officials were most courteous and prompt; I there found the old War Office papers, but alas! as regards Berkshire, they were somewhat scanty. After this I returned to my search nearer home, in our splendid Free Library at Reading. The following pages are the result of my various researches and enquiries, and I feel sure they will prove of interest to past, present, and future Berkshire Militiamen.

There are already written histories of several Militia Regiments: the 2nd Royal Surrey, the Cambridgeshire, the Bedfordshire, the Northampton and Rutland, the Hampshire, etc., etc. All these give the origin of the English Militia from Saxon times, with plentiful extracts from Grose's *Military Antiquities*, Palgrave, Blackstone, and the *State Papers*, and, though all this is most interesting, I shall try to treat the subject, as far as possible, as entirely "Berkshire." But all Militia histories must bear a strong family likeness, as they can only be compiled from the few existing MSS. and books on the subject.

To write any history thoroughly would mean the work of a lifetime, to the exclusion of all other occupations. My life is far too busy for me to give up more than a few spare moments for any literary work; even while writing this short preface, I have been interrupted half-a-dozen times. Also as regards research, I have not been able to go thoroughly into the matter for the same reason ; but if my readers will be so good as to send me any notes relating to the regiment or its officers, it will be of great use and value for future editions.

It has been a great pleasure to me, searching out and chronicling the exploits of the Berkshire Militia, in which my father served from its reorganisation in 1852, until a serious illness, of two years' duration, compelled him to resign his commission in 1873.

From my earliest childhood I can remember the regiment; and the study of its past history, I am glad to say, is one long record of praise, without containing a single blot or blemish, throughout the many eventful years which have elapsed since 1640, the earliest date at which I have found the term Militia used in reference to the regiment.

This regiment was generally called the "Berkshire Regiment of Militia," often abbreviated to "The Berkshire Regiment." This appellation is claimed for the 66th Regiment, but the latter regiment was only raised in 1759, and consequently is comparatively modern to the Berkshire Militia, which also enjoyed the privilege of being a "Royal" regiment more than seventy years before the present Depôt system of Linked Battalions came into existence.

EMMA ELIZABETH THOYTS.

*Sulhamstead Park,*
*1897.*

A 2

# INTRODUCTION

BY

COLONEL BOWLES.

———

IN dealing with the History of any Militia Regiment, it would seem advisable shortly to touch upon the question of what is meant by "Militia" generally, and to · trace in outline its existence in England from the earliest known times ; chiefly because the ignorance, not only of the general public, but even of those connected with the Force appears to be profound.

All over the known world from the earliest times of the formation of men into societies, tribes, or nations, there seems to have existed an acknowledgment of the duty upon all males of suitable age, of bearing arms in defence of their own particular society, "their hearths and homes," under their local or tribal leaders, chiefs of tribes, or septs, officers appointed by their kings, or paramount chiefs, or in some cases by elected leaders ; and in the earliest times it was recognised that this duty, incidental to their citizenship, or membership of their own society, tribe, or nation, should be performed not only without pay, or reward, save such as they took from their enemies, but themselves finding their own arms and accoutrements. This first prin ciple is after all a sound one and of easy comprehension. and though of necessity modified as a more complex system of society grew up and the Art of War became scientific, yet it underlies at the present day our own Militia system, and is recognised by the law of ballot for our Militia, which, though now in abeyance, is the law of the land.

This theory is more evidently adopted by all the great nations of the Continent of Europe, as will be perceived on consideration, with their universal liability to compulsory service; for, of the prodigious forces available in their armies, though they keep up always the cadres and organization of the whole, yet only a part, are so to speak, in training at a time; the remainder being held in readiness to be called up to their places in the first line, or to form, according to their age and qualifications, a second and third line. It may almost be said then that they form an enormous National Militia, but with part permanently embodied and the organisation of the whole kept complete.

England is in a somewhat different position, chiefly owing to her Foreign possessions, to India, and her Colonies. To a certain extent also, this applies to France, who in a sense has a separate Colonial army, at any rate in Algeria. In our own case then it is manifest that the same system as that pursued in other countries, based, as I think we may say, upon the Militia theory, would not do by itself, for it has always been part and parcel of that theory that the Militia should not, at least in time of peace, be called upon to serve out of their own country. It has been found absolutely necessary therefore, for England to have an Army, a standing Army of another kind, one recruited entirely by voluntary enlistment, to meet the exigencies of all our foreign service.

It is, however, the opinion of many thinking men that we are in danger, from not sufficiently recognising in these modern days what I have called the Militia theory, of some day courting disaster. Our standing army is not a large one, barely in fact large enough for all it might at any time be called upon to do out of the country, and it must be remembered that though in times past over and over again, in National emergencies, the country, in its need has fallen back upon its Militia Forces, neglected in time of peace; yet, that now when things move so much more quickly, and means of communication are so much greater and speedier, when with the present arms of precision a soldier wants so much more training and our possible opponents are so much better trained and equipped, a large addition to our Militia or defensive forces cannot by any possibility, however great the patriotism of the moment, be organised and trained at a moment's notice.

Of course, a compulsory Militia service is in the nature of a tax upon the country. England, however, is perhaps the most lightly taxed country in the world, as well as the richest, and such a tax is after all only in the form of an insurance on our riches, and one which, I believe, the people of this country would be content to pay, if convinced of its necessity before the day comes when it may be too late, and when all our frantic endeavours and expenditure at the last moment may be unavailing. I believe, also, that the individual benefit of the Military training to our people would be great. I will only add that it is the opinion of those best able to judge that the natural defence given by our insular position is not so great as once it was.

To turn to the Militia question in this country in earliest times and thenceforward, we know that it was the custom of the Romans to turn what form of Militia or fighting power they found in conquered tribes or nations to their own use, calling on the tribes to find a quota of men for Militia service, in addition to such men as went to serve permanently in some of their legions; no doubt, therefore, they pursued some such policy during their occupation of Britain.

In Saxon times, however, we find under the various Kings of the Heptarchy a very complete Militia, under the name of the Fyrd, and fines were imposed on those who shirked the duty, in addition to the scorn with which they were regarded by their neighbours. It may be interesting to note that the word " Fyrd " or " Frid " is an Anglo-Saxon word, with the double meaning of peace and freedom, and " Frid-gegild," a protection society. When the Heptarchy, after continuous wars among themselves, came to an end, and the Kings of Wessex became Kings of all England, Alfred the Great, in danger continually from the Danish invaders, reorganised the Fyrd and made it a very valuable defensive force, defeating again and again the Danes with it.

This is of special interest to men of Berkshire, for Alfred was a Berkshire man, and doubtless many a Berkshire Militiaman of those days served in his defensive wars and in the battles fought in the county from Reading westwards.

The Fyrd seems to have been kept in a fairly efficient state for long after Alfred's time, and, indeed, down to the Norman Conquest, fighting with all their Saxon obstinacy under Harold at the Battle of Hastings, and though opposed to the better armed chivalry and the greater warlike genius of the Normans, yet giving the invaders hard work to defeat them, and dying in enormous numbers with their faces to the foe, and all their wounds in front, around the last Saxon king.

With the advent of the Normans was introduced the feudal system, into which it is not necessary to enter here, but side by side with it the shrewd Norman monarchs kept up the Militia system to a great extent, doubtless finding in it as time went on a counterpoise to the power of the great barons. We find from time to time statutes and ordinances for the better mustering, arming and regulation of the local forces, which it may not be out of place briefly to refer to here, though many of the later ones are dealt with at more length in the body of this work.

To go back to King Alfred's time, about the year 879, when by the aid of the local forces of the Fyrd he had established the Saxon kingdom, we read in Grose's *Military Antiquities* that every ten families made a tything, and all the males of those families of military age were commanded by the borsholder, called conductor. Ten tythings constituted a hundred ; the soldiers of the hundred were led by the chief magistrate of the hundred, sometimes called the hundredary. Several hundreds formed a trything, commanded by an officer called a trything man ; the force of the whole county was commanded by the dux, or duke, in peace time, and by the king himself or his appointed officer in time of war. The hundred elected their own officer, and, it would seem, acknowledged his authority by touching his spear with theirs. Up to the early part of this century the hundred was still recognised as a Military division of the county, the tything man's duties having devolved on the parish constables. In Saxon times the penalty for not serving was, in the case of owners of land, forfeiture and fines of varying degree; at the beginning of Alfred's reign the fine seems to have been sixty shillings for a landowner and thirty for a churl, a large sum in those days.

After the Conquest, as we have seen, the Fyrd, later on called the *posse comitatus*, was still continued by the Norman kings in addition to or as supplementary to the feudal system. In it all men between fifteen and sixty were liable to serve in their own county, and, in case of emergency, anywhere in the kingdom. Once a year there was a review of arms in each county; every owner of land had to provide armour according to his possessions. The *posse comitatus* was placed under the sheriff, who had to keep the king's peace, and times were fixed for instruction in arms.

In the Assize of Arms of Henry II., 1181, all the freeholders and burgesses were bound to provide themselves with armour and weapons according to their degree, so that they not only had to serve, but to find their own offensive and defensive arms; the arms to be found by the different degrees will be found in Grose's *Military Antiquities*, vol. i. No one could sell, give away, or otherwise dispose of his arms, neither could a lord seize his vassals' arms.

Then came, in 1285, Edward I.'s Statute of Winchester, much on the same lines, but with somewhat different provisoes; all were by this statute (13, Ed. I., c. 6) ordered to produce their arms twice a year on penalty of being " presented " by the constables of the hundred.

In 1553, arms and other circumstances having altered, the Statute of Winchester was repealed, in the reign of Philip and Mary, and fresh regulations enacted in place of the old ones ; still, arms and armour had to be provided in the different degrees until the reign of James I., when the providing of armour was abolished. The richer men under these enactments had to provide horses and accoutrements for mounted men.

In the reign of Henry IV. we find certain persons called Commissioners of Array, a term synonymous with the more modern Lords-Lieutenant, who had to muster the men and issue the County Muster Rolls, which will hereafter be found alluded to. In the reign of Henry VIII. the Lords-Lieutenant had succeeded to their duties, and the command of the Militia was vested in them under the sovereign on two occasions, in 1558 and during the Long Parliament, and they had power given them to press men for the Militia; but, in each case, it was specially enacted that these men should not be compelled to go out of the country. No doubt this proviso was inserted for Constitutional reasons, for some of our kings had endeavoured to assert their right to take the men for foreign wars, though, when this was done, they seem to have received pay.

In the early part of the reign of James I., when the provisions as to finding armour were repealed, the command of the Militia was taken from the Lords-Lieutenant; and in the reign of his son, Charles I., the dispute as to the control of the Militia, which the Parliament wished to usurp, was one of the principal causes of the Civil War.

We now approach the time of the first formation of any Standing Army, unless we look upon the feudal retainers, the Knights of the Hospital and various bands of mercenary troops from time to time employed, in that light. The feudal system had disappeared; and I think it may be said that, up to 1660, the Militia was the only Constitutional force of the country. Even long after this the people looked somewhat askance on a Standing Army, as putting too much power into the hands of the Crown.

On the Accession of Charles II., the King was confirmed in control of the Militia, with the Lords-Lieutenant under him in their several counties, and times for muster and training were appointed, once a year for complete regiments and four times a year for single companies, but not to exceed twelve or fourteen days in all.

Still the people had to provide arms and accoutrements, and, in some cases, horses for mounted men, according to their degrees; but a rate was levied for the supply of ammunition, drums and colours.

The inadequate time for training, however, and the money spent and attention given to the Standing Army caused, by degrees, the Militia to become very much neglected, and in some counties even the musters were made very irregularly.

This went on till 1756, when, there being a fear of invasion, the King (George II.) proposed to bring over some Hanoverian troops. This the people could not stand, and the attention of the Government having been already for some time previous called to the state of the Militia, the force was then reconstituted much in the form in which it existed down to the time of the modern territorial system, when it was taken out of the hands of the Lords-Lieutenant and linked with the Line battalions, under territorial designations; the officers, at the same time, being placed permanently under the Army Discipline Act. One curious effect of the new organisation is that no one knows to whom it would fall to enforce and carry out the Ballot, if required; the old power of the Lords-Lieutenant having been taken away.

Most of the Militia regiments now in existence date, in their present form, from about 1756 or later.

This summary would hardly be complete without some slight allusion to the martial regulations, now represented by the Army Discipline Act, under which, from time to time, the force has served. The earliest of which there is direct record is that of King John, entitled, "Constitutions to be made in the Army of our Lord the King." Next, those of Richard II., twenty-six in number (Grose's *Military Antiquities.* vol. 2). Of Henry V., made chiefly for his feudal Army abroad. Of Henry VII. and of Henry VIII., 1513. These last form the basis of the modern Acts. Then we have those drawn up for the King's forces and also those for the Parliamentary forces, in the Civil War, which were very severe, according to the customs of those days, though, by the old Court-martial Books of the Royal Berkshire Regiment, it appears that, even at the beginning of this century, tremendously severe floggings were imposed and actually inflicted, as many as four, five and six hundred lashes being not infrequently the sentence, for what we should now-a-days inflict a few hours' imprisonment with hard labour. The last of the old ordinances connecting modern times with old are those of James II., more lenient than the old ones; they were called the Articles of War, were sixty-four in number and under them no one was to be punished during peace time by loss of life or limb, though such punishments were authorised in time of war.

Finally, to summarise the services of the Militia forces of the country, without again referring to the days of Alfred or those of Harold and the Conquest, they took part in all the internal troubles and invasions of the feudal times; in putting down the various insurrections; in the long Wars of the Roses; in the preparations to resist the Spanish Armada; and on both sides, often in duplicate as it were, in the great Civil War; at the Restoration; at the time of Monmouth's Rebellion; in the "15" and the "45;" almost continuously from 1760 to 1816; and, finally, during the Crimea, and Indian Mutiny time, both at home and abroad.

May we hope that when our Queen and country want the services of the Militia Battalions in the future, they may be found with the men, the discipline and organisation, to enable them to perform such services as they have in the past. Given the means and the opportunity, I feel sure they will not fail the country at need.

T. J. BOWLES.

# CONTENTS.

|  | PAGE. |
|---|---|
| PREFACE ... ... ... ... ... ... ... | v. |
| INTRODUCTION ... ... ... ... ... ... | ix. |

CHAPTER I.

| MILITIA AND BERKSHIRE LEVIES ... ... ... ... | 1 |

CHAPTER II.

| EARLY ENTRIES OF SOLDIERY, 17th CENTURY ... ... | 16 |

CHAPTER III.

| THE CIVIL WAR—1640-1649 ... ... ... ... | 25 |

CHAPTER IV.

| UNDER THE COMMONWEALTH—1650-1659 ... ... | 42 |

CHAPTER V.

| RETURN OF THE KING—1660-1715 ... ... ... | 51 |

CHAPTER VI.

| THE HANOVERIAN DYNASTY—1715-1757 .. ... ... | 63 |

CHAPTER VII.

| REORGANIZATION AND EMBODIMENT—1757-1763 ... .. | 75 |

CHAPTER VIII.

| WHEN GEORGE III. WAS KING—1764-1792 ... ... | 93 |

CHAPTER IX.

| WARS AND RUMOURS OF WARS—1792-1803 ... ... | 122 |

CHAPTER X.

| NINETEENTH CENTURY WARFARE—1803-1852 ... ... | 157 |

CHAPTER XI.

| CHANGES AND IMPROVEMENTS—1852-1872 ... ... | 177 |

# Contents.

CHAPTER XII.

TWENTY-FIVE YEARS IN THE REGIMENT—1873-1897 ... 191

CHAPTER XIII.

PLACES WHERE THE REGIMENT HAS BEEN—1614-1896 ... 232

CHAPTER XIV.

LORDS-LIEUTENANT OF BERKS, AND COLONELS OF THE REGIMENT—1640-1897 ... ... ... ... ... 245

CHAPTER XV.

OFFICERS WHO HAVE SERVED IN THE BERKSHIRE MILITIA 248

AUTHORITIES QUOTED FROM ... ... ... ... 333

INDEX OF NAMES ... ... ... ... ... ... 336

INDEX OF PLACES ... ... ... ... ... ... 344

INDEX OF REGIMENTS ... ... ... ... ... 348

OFFICERS IN 1897 ... ... ... ... ... 351

## LIST OF ILLUSTRATIONS.

|                                         |           | PAGE.         |
| PORTRAIT OF COLONEL BOWLES ... ... | .. to face Title. |
| —— COLONEL WHICHCOTE ... ... ... | 43 |
| —— JOSEPH ANDREWS ... ... ... ... | 76 |
| —— JOHN WILDER ... ... ... ... | 96 |
| THE MARCH ... ... ... ... ... ... ... | 126 |
| PORTRAIT OF COLONEL LOVEDEN ... ... ... ... | 128 |
| THE COLOURS OF THE REGIMENT ... ... ... ... | 181 |
| CORFU AND VIDO ... ... ... ... ... ... | 186 |
| THE PARODY ... ... ... ... ... ... | 189 |
| REGIMENTAL PLATE ... ... ... ... ... | 190 |
| THE REGIMENT ON PARADE ... ... ... ... | 204 |
| GROUP OF OFFICERS .. ... ... ... ... | 226 |
| COLOUR SERGEANTS ... ... ... ... ... | 230 |

# THE
# ROYAL BERKSHIRE MILITIA.

IES.

land which have
y than Berkshire
has always done from the earliest ages until
modern times, and the Berkshire Militia may
fairly boast they possess a longer consecutive
history than any other Militia regiment, for
they were recognised as a Militia regiment from 1640, and
their longest period of inactivity was for some twenty years
prior to 1852.

For the origin of the Militia it seems to me more probable
to search back to the Roman Period, when
200 A.D. England was under military rule. We know that
the Romans recruited their armies from every country they
conquered, consequently many Britons must have served
under their standards both at home and abroad.

Curiously, Berkshire possesses few relics of the Roman
occupation. In this county no large town has ever been
found, the sites of camps are pointed out and many large
villas, although the nearness of Berkshire to London, and
the fact of the River Thames flowing through it, must have
made it an important district in all ages.

B

Both Saxons and Danes fought here.   King Alfred's
birthplace is said to have been at Wantage.   This king, so
popular in history, has the credit of organising the Militia or
"Fryd," a statement for which the antiquary Grose is mainly
responsible.   The Saxons soon adopted England as their
home; they lived in tribes or settlements, and that they
possessed some system of raising troops on emergency was
inevitable.   Perhaps King Alfred found some arrangement
existing, which he reorganised on fixed rules and
900.    stated lines.   From then to the Norman Conquest
such matters are vague.   We can get no information, as the
few writers of the period were chiefly ecclesiastics, who dealt
with history from their own point of view, and did not
trouble about military affairs.   In the Battle of Hastings,
the English Militia, though brave in the field, were conquered
by the superior military skill of the Norman troops,
1066.    for a disciplined body of men can always gain the
day over masses of raw recruits, however superior they may
be numerically.   Glancing through the pages of English
History, we see a long succession of wars, in all these
Berkshire men took part.   First comes the Empress Maude's
gallant struggle to secure her father's throne.   That king, who
had founded Reading Abbey, choosing it for his last resting
place, must have been well known by the men of Berkshire;
we may be sure they rallied round his daughter
1139.    and supported her interest.
After this, for several generations, the Crusades excited
the military feelings of the whole then known world.   Many
Englishmen perished in these Holy Wars; there is no
means of ascertaining the number of the soldiers who went
out from England at that time.   Then King John and his
Barons disagreed, and all the 12th, 13th and 14th Centuries
were full of fighting at home and abroad, till a long civil
war, the war of the houses of York and Lancaster, distracted
England.

"Arrays" and "Musters" are not fully entered into in history. They were ordinary matters of no interest and must have taken place periodically, as we know that in every war the troops who fought were raised by levy from the towns and counties of England, equipped and paid for by special taxes levied for the purpose all over the country. The men were retainers or followers of whoever lead them to battle; they wore special liveries or uniform to distinguish them; these liveries were originated from the heraldic device or coat of arms of the leader, which was borne on the standard. Scarlet has always been the English colour—except, perhaps, in the Tudor days, when it is said the royal livery was green and white. Red, white and blue, were the colours of Mary, Queen of Scots, chosen from the combined coats of arms of England, Scotland, and France.

Our Union Flag, called the Queen's Colour, bears the crosses of St. George, St Andrew, and St. Patrick, as representing all parts of the United Kingdom, it was re-arranged in 1800, when the Union was declared; previous to that date it bore only the two crosses.

I do not trouble to seek every notice of military affairs during the earliest days of history, indeed, it would not be possible to find many entries of local interest; I confine myself strictly to those only in which Berkshire is distinctly mentioned. To search for such would be a long and expensive matter. The "Arrays" developed into the "Musters" and "Views" (query, is our word Review taken from this obsolete word?); these in turn became "Trained Bands," though the "Trained Bands" were chiefly town troops, who had more opportunity of drilling than the Militia recruit from the county. The word Militia, then spelt "Milicia," is found as early as the reign of Queen Elizabeth, but it did not come into general use until the Restoration of Charles II., when the formation of a standing Army began to be thought of, and new regiments of regularly trained

B 2

soldiers were embodied permanently, and recruited from all over England.

But under whatever name they might be called, the Militias are, undoubtedly, the oldest corps in England and are justly proud of the fact. Their origin goes back hundreds of years before any Line Regiment was thought of. Previous to the 17th Century, the Nation depended entirely upon them for defence, they fought both at home and abroad, and, though there remains but scant information as to individual regiments, we know that then as now "England expected every man to do his duty," and she could not have occupied and held the high position she did among other Nations, unless her soldiers had been loyal and brave.

In the second year of the reign of Edward III, at a meeting of the Town Council of Reading, before 1462. Thomas Beke, Mayor, Friday (November 17th) next before the feast of St. Clement, there was reserved for the hands of Stephen Donster, 5s. 10d.; to be paid for arrows (arrys), 6d.; and for the Mayor, Thomas Beke, of his accounts had for soldiers that last went to the King, 19d.; also for Robert Stapper the same day, 3s., owing, 2s. These men had advanced the money which was to be repaid them out of special rates leived for the purpose.

This was in the War of the Roses, Edward IV. being crowned at Westminster after the battle of Mortimer's Cross in 1461, but fighting continued for many years after.

A Parliament was held at Reading in 1451, and again in 1452, owing to an outbreak of the plague in London.

Edward IV. also held Parliament there in 1466, which was probably the last Parliament ever held in the town, for though Henry VII. and Henry VIII. both visited it, no mention is made of any councils held, and the 1466. splendid Abbey where kings and princes had been royally entertained ever since its foundation, ceased to exist in 1539; and Charles I., when he came hither was without a Parliament, his policy being to govern the nation by the Divine right of kings, without interference from Lords or Commons.

The Battle of Barnet, 1471, was quickly followed by the death of Henry VI., this, and the imprisonment of Margaret of Anjou, at last terminated in 1485 that long and terrible war which had affected England. The fighting in these wars was in the middle and northern parts of England: that portion of the country which lay between York, the northern capital, and London, the capital of the South of England ; but the soldiers who formed the two armies were levied from all parts. Berkshire probably sided with the Lancastrian party. The Duke of Suffolk, a prominent leader, married Alice, daughter of Sir Thomas Chaucer,

1471. and from her he acquired much Berkshire property. William de la Pole, Duke of Suffolk, had been instrumental in arranging the marriage of Marguerite of Anjou, he was popular, not only as a courtier, but more so as the General of the Army in France ; but his popularity ceased when ill fortune attended his army, and his tragic death by beheading in a tiny boat in the midst of the English Channel is one of the most extraordinary episodes to be found in history.

August 22nd. Henry VII. An Inventory was 1488. made of the Armour in Reading. Four pairs of brigantines, two covered with russet fustian and two with black fustian. (This was armour composed of small pieces of metal sewn on leather and covered with cloth.)

Another pair of brigantines covered with black : a jack ; four sallets. (These were the head pieces.) Two pairs of gossets ; two aprons ; three standards ; one pair of splints ; a black bill ; two sheafs in the bundel gyrdeled (girdled). (This consisted of forty-eight arrows, as a sheaf contained twenty-four arrows).

Reading was certainly not well provided, if this was all they had in the town ; as half-a-dozen soldiers could not do much for the defence of a garrison. But according to the Statute of Winchester passed in 1285, every householder was obliged to keep armour, so that in cases of need, each town would have been defended by its inhabitants, each armed at their own expense, and the weapons, etc., belonging to the Corporation only appertained to its own special guard of soldiers.

1488.

A Tax of £7 was levied to provide harness for six men in Reading. This again evidently refers to the Town Guard, whose equipment the previous year had been examined and was far from complete.

1489.

War with France was begun by the English King, who pledged himself to aid the Duchess Anne of Bretagne, a defenceless girl of twelve years old, to protect her kingdom against the claims of the French King. One of the terms on which this aid was sent, was that Anne of Bretagne should not marry without Henry VIII.'s consent, but Charles VIII. of France forced her to marry him, despite her bethrothal to Maximilian, King of the Romans. It needed but slight excuse in those days to create war between England and France, so in 1492, Henry laid siege to Boulogne, but the English were impoverished by the long wars of the 15th Century, and the King soon after wisely secured a treaty of peace, together with a large sum of money. The next event in English warfare of interest was the pretension of Perkin Warbeck, the "White Rose of York," to the Crown. His first attempt ended in defeat near Deal. From thence he went to Ireland and Scotland, (the Scotch King finding him arms and money), and led an army over the border to support his claim. In Cornwall there was a rising in his favour, and the Insurgents marched to Salisbury. But though Perkin had to fly to Ireland, he did not despair. His next venture was in the West of

England; again at the most critical point his courage
1496. failed, he left his army and fled, and his chance of
a Crown was gone for ever, for he was taken prisoner.

Twenty-nine soldiers were provided by the Town of
Reading for the King. Nine of them were
1514. bowmen, the weapons of the rest are not mentioned. Each town was taxed in proportion to its wealth
and population. It is not known on what terms the soldiers
served, or whether at the end of a war they were sent home
again ; possibly few survivors lived to return.

County districts as well as towns had to provide soldiers.
Thus if Reading sent 29 men, and the parts were taxed in
proportion, several companies must have been raised, over
and above those kept for local defence.

Soldiers were sent from Reading to Scotland, 12th September, when Henry VIII. and James of Scotland were
1542. to have met at York, but the latter did not come, a
breach of promise which Henry made an excuse for warfare.
In August, 1542, the English troops crossed the border.
Two months later, in October, Henry issued a fresh
Manifesto claiming the sovereignty of Scotland, the real
object of his quarrel with the Scotch King, being to annex
that kingdom to the Crown of England. Twenty-four
horses and their harness and a soldier to each, were sent by
inhabitants of Reading to assist the King. The names of the
soldiers were :—John Seagrove, Lewis Baker, John Taff, John
Gateley, Nicholas Norres, John Withwall, Richard Ryce,
Thomas Evett, William Coker, Moses Cutler, William Rows,
William Mayle, Thomas Alyn, Henry Freman, Richard Est,
Richard Hensman, Matthew Hoskyns, Nicholas Gent, John
Dole, Roger Statham, John Cordery, John Hopton, William
Pulleyn, John White.

All the inhabitants were taxed to provide this troop, both
Clerical and Lay.

Some authorities say Henry VIII. gave power over the Militia to the Lord Lieutenants of Counties. Others assert it was not granted them till Queen Elizabeth's reign. Perhaps both statements have foundation in the fact that the Lord Lieutenancy gradually grew into a more responsible office.

May 12th. The Town of Reading sent to the King, to
1544. help in the French wars, thirteen men well harnessed and horsed, and twenty foot men well harnessed at the charge of the said town. They cost about 40 marks to equip. The Duke of Suffolk, a King's Lieutenant and Captain General summoned all captains, vice-captains, men-at-arms, armed men, archers, and others of the counties under him, namely—Berks, Kent, Surrey, Sussex, Middlesex, Bucks, Hants, Wilts, Oxford, Worcester, Hereford and London. The siege of Boulogne was the chief incident of this war, after the success of which Henry returned home.

Shortly after, Francis raised the French Fleet and started to conquer England. The two navies met in the Channel, but after a cannonade of two days, the French returned home, finding England was too fully prepared for war to fall into their hands.

The Privy Council orders of Henry VIII. as given to the
1544-5. Duke of Suffolk, as the King's Lieutenant and Captain General of the Forces, to raise troops to carry on the war with France, are minute and curious. The first thing ordered relating to musters, was to ascertain the number of able men in each hundred, and "how many of them were archers, how many billmen, how many furnished with harness, bowes and arrowes, and weapons convenient to serve in case of need," and care was to be taken that men, horses and weapons were ready to be used "upon one hour's warning," not only for the defence of that immediate part of the country, but elsewhere, where the King might require them. The coasts were to be examined.

Any places likely to afford a safe landing to invaders were to be repaired and fortified by trenches and earthworks, beacons being set up near at hand to give warning of the approach of an enemy. The dwellers near the sea coasts were to take notice if they saw any number of ships hovering about searching to land, and if the landing were effected, the country people were to break the bridges, and cast up trenches to prevent a further progress inland. Watches were set in towns and villages along the coast ready to give the alarm and muster the soldiers. It was the business of the Lieutenant of the Shire to give notice to the Justices of the Peace, and instructions were issued to them for this purpose. It is very clear in reading of these warlike preparations, that Henry VIII. momentarily expected an invasion from the French, if he failed in the bold attempt to reduce that nation himself; for in the expedition thither he took with him the bulk of the English Army, so that this country was entirely dependent for defence on the careful carrying out of the orders for musters. For the training of the men, if not carried out by the gentlemen of the neighbour-hood, the Lieutenant was to appoint from the King, certain chosen captains for the purpose, "To teach and train the people how to wear their arms and use their weapons." This instruction being most conveniently arranged for the afternoons of holy days at some given spot, for two or three months' space. Even the prices of the ammunition are given, so the Lieutenant had only to carry out his orders, and similar letters of instruction were issued in the succeeding reigns of Edward VI., Queen Mary, and Queen Elizabeth, becoming more and more detailed and explicit.

February 4th. A certain captain was appointed by the 1550. King's Majesty, and by his honourable Council to take within the Borough of Reading, fifty soldiers to be weaponed with bills, swords, daggers, and certain bows and sheaves of arrows at the cost and charge of the

inhabitants, to every soldier 12d. in money, over and besides
their coats and pressed money at the King's charge. This
was probably the first payment of bounty or conduct money.
Hitherto the soldiers had been more of the nature of
retainers raised on the feudal system of forced labour, than
hired servants. The direct payment from the Crown was
the step towards a regular army, as previously the men
had been clothed and paid out of taxes levied for the
purpose.

July 21st. The Mayor and inhabitants of Reading sent
out of the Borough of Reading to aid the Queen
against the Duke of Northumberland, ten soldiers
well harnessed and weaponed, at the expense of the Mayor
and inhabitants. Troops were being raised and sent to
quell the fatal Rebellion, which ended in the execution of
poor Lady Jane Grey. The prompt measures of Queen
Mary's Council and their unanimous declaration in her
favour, frightened the Duke of Northumberland. He
declared for Queen Mary, but this did not save his life, for he
was taken prisoner, his army disbanded, and himself conveyed
to the Tower where he was executed, August 22nd, 1553.

On 16th August, the Mayor and inhabitants of Reading
sent six men well harnessed, well horsed, and well apparelled,
to attend and wait on the Queen's Majesty at Richmond
for her Coronation. The men returned to Reading, August
25th, when the ceremony was over.

This is the only occasion on which I find a notice of
Berkshire men taking part in London ceremonies. London
was better provided with soldiers than any part of the
Kingdom, yet it is curious, that as Windsor was a Royal
residence, the County should not have been called upon to
find a Royal Body Guard on state occasions, such as
Coronations or birthday processions.

July 9th. The Reading people sent to King Philip and
Queen Mary forty men in blue coats with red crosses, which

1556-8. cost 6s. 4d. each man. They had forty new bills, each worth 18d. and conduct money 16d. per man. The total amounting to £18 6s. 8d. which the inhabitants had to pay.

This allusion to uniform is interesting, as little if any information is obtainable as to the uniforms, or livery, of the Trained Bands, though each Troop was distinguished by a special colour. The Red Cross Knight of the old song at once comes to one's mind :—

" And on his breast a bloodie cross he bore,
    The deare remembrance of his dying Lord,
  For whose sweete sake that glorious badge he wore,
    And dead as living ever, him ador'd
    Upon his shield, the like was also scor'd."

Spencer's Faerie Queen.

The Red Cross was undoubtedly the badge of St. George for England, our Patron Saint, which forms so conspicuous a part of the Union Flag.

No doubt in the middle ages the different troops bore the livery or colours of their leaders. Each officer 1558. could then easily pick out his own men by their dress, and was responsible for all who wore his colours. Our servants' liveries of the present day are remains of this custom.

The war with France was declared June 7th, 1557. The Queen levied 1,000 horses and 4,000 foot soldiers, 1557. and 2,000 pioneers, and sent them to Flanders in July, under the command of the Earl of Pembroke, with Lord Robert Dudley as master of the Ordnance. The English Army joined the Foreign Troops of King Philip's Army under the Duke of Savoy.

The war terminated in the loss of Calais, which it is said broke Queen Mary's heart ; hers was a sad loveless life, from cradle to tomb. By historians she has been blamed, not pitied, and her name handed down to posterity with ignominy.  Now looking back, without any of the prejudices of contemporaries, we see clearly the trials and difficulties surrounding her on every side.  Her death took place November 17th, 1558, after 'a reign of five years and a few months.

1558.

Formerly in the Parish Register of Cholsey, there was a curious loose paper, since lost.  It contained instructions to the young men of the parish to practice the long bow twice a week, at Butts on a piece of ground set apart for the purpose.

The Butts in Reading are so called to this day, but it must be borne in mind that the word " Butts " was also used for a measure of land.  Perhaps the archery grounds were called " Butts " from being a certain measurement. The old yew trees in churchyards were, it is said, planted by royal command, to ensure a supply of yew timber, for the making of bows.

May 19th.  At a meeting of the Burgesses of Reading, they referred to Thomas Bygges old account, and the account and reckoning for sending soldiers to St. Quentins, Robert Tylby, Richard Gilbert, Walter Berrington, and William Lyppescombe having stood surety for the money expended, which had not been repaid to them. Walter Berrington and Robert Tylby are named among the cofferers of Reading.  Their names appear too in the list of Burgesses as does also William Lyppescombe, but not Richard Gilbert.

1559.

With Queen Elizabeth a new era began.  She was unmarried, free from foreign alliance, with clever advisers to direct and counsel her.  Her subjects were weary of religious persecutions, they desired peace at home and abroad, and all Europe was anxious to be friendly with the Queen, who, it was generally supposed, would seek a husband from one of the Courts of Europe, as her ancestors and predecessors on the English Throne had done.

The reign of " Good Queen Bess " is remarkable rather
for diplomacy than war. She ruled by stratagem instead
of military force, but none the less she was ready for war or
defence when necessary. Martial Law was em-
1559. ployed in those days against civilians as well as
soldiers; its method was prompt and severe. The Sovereign
had absolute authority, which none dare question, and she
used her prerogative.

A General Muster was ordered ; this shewed
1574-5. the militiary resources of the country, and the
Militia of England was at this time established at 182,929
men.

This army, although large, was undisciplined. The men
seldom met for drill and were unaccustomed to moving in
bodies. At this time Lord Lieutenants of counties were
placed at the head, and practically in command, of the
Militia.

In this Muster it is said that the men fit for service
amounted to 172,674, but it was believed this did not give
the full number, as the returns were imperfect. Queen
Elizabeth by calling this Muster was endeavouring to obtain
an estimate of her available forces, and organize her army,
which, without doubt, required much reform and discipline.
The gathering together of the countrymen as soldiers
taught many valuable lessons ; furthermore, it
1575. restored the confidence of the country, and avoided
the panic consequent upon threatened invasion. Thus,
some years later, the Queen was able to organise a line of
defence all along the coast of England, when the long
deferred invasion of the Spanish Armada became imminent.

The troops assembled at Tilbury, some 22,000
1585. in number. The Berkshire " footmen " were 1,000
strong ; the Light Horse of Berkshire numbered 200.

The army to guard Her Majesty's person consisted both
of horse and foot soldiers. This was under the charge of

the Lord Chamberlain.   The Berkshire contingent of the
Queen's bodyguard was 230 horsemen.  The Queen's speech
to the troops inspired them with loyalty, they were ready to
defend their country and their Queen against any foes.
Even if the Spanish ships had conquered the English fleet,
they would have found it no easy matter to effect a landing
on the coast, despite the want of military training which
must always mar an army hastily gathered together for
defence ; but the English troops then raised had been pre-
paring for many years for home defence.   Queen Elizabeth
was too wise a sovereign not to have attended to such
matters.   The Musters of 1574 shewed the resources of the
counties.   Although not mentioned in history, we may

1586.   safely assume that annual exercise took place, very
much as that of our own Militia at the present day.

1588.   Queen Elizabeth ordered a Special Muster of the
Troops of each county to guard the coasts during
the threatened Spanish invasion.   The Lords Lieutenants
sent in returns of the Militia forces, and without any troops
from London, the number amounted to 130,000 men, whom
a little drill would soon have made into good soldiers.

Two thousand foot and 200 horse were sent to Milford
Haven, 5,000 Cornish and Devon men guarded Plymouth.
From Dorsetshire and Wiltshire, Portland was protected,
and at all other important points of the coast were soldiers
quartered.   The celebrated Review at Tilbury, when the
Queen in person inspected her Army, took place in July ;
the arrival of the Spanish Armada was then hourly expected.
The delays, misfortunes, and ultimate annihilation of that
splendid navy are graphically described by Froude.   Our
soldiers from the English coast must have watched with
anxiety the movement of the Fleets.   When the great ships
had gone, the Camp at Tilbury was disbanded in August,
although the Captains and Officers were not discharged
until later, when all fear of further invasion was at an end.

1599. One hundred and forty men were furnished by Reading, August 9th, against the intended invasion of the Spaniards. The English Army being in Spain under the Earl of Essex, the Spanish thought this a good opportunity to revenge their previous defeats, but this expedition ended without any result. On the other hand after some success in Spain, the English Fleet and Army returned home, and Philip of Spain died soon after.

As a Military Leader, the Earl of Essex was beloved by his soldiers, he was a better soldier than politician. It was sad that so brave a commander should fall a victim to political intrigue and suffer death as a traitor. On Ash Wednesday, 25th February, Robert Devereux, Earl of Essex, was beheaded in the Tower. He was only 33.

The long reign of the Virgin Queen was nearly over ; the death of her favourite Essex preyed on her mind. She had ruled her people wisely and well for many long years. It was best she should die before her powerful intellect failed.

# CHAPTER II.

### EARLY ENTRIES OF SOLDIERY.—17th CENTURY.

IT is most difficult to work out any Military matters during the early part of the 17th Century. Historians are silent on the subject, save a few notes here and there. Regiments were then more of the nature of Private Troops, called by their Commander's name. Though raised or called out by the Lord Lieutenant's, or Deputy Lieutenant's Warrant, only the King appeared to have any actual authority over them; indeed, James I. took away the power previously granted to the Lord Lieutenant, and reserved to himself absolute control over the soldiers of England.

The struggle between the King and the Commons was smouldering for years, the billeting of soldiers in private houses, the tax called Ship money—although it was partly expended on the Army—and, finally, the question as to whom belonged the right of calling out the Militia, all purely military matters, were looked upon as grievances by the people, who dreaded a standing army. Soldiers were "pressed" from the lowest of the people, "idle persons," as some of the records quaintly call them, and military discipline and organization, though severe when administered, was faulty, not being fixed by law or Act of Parliament. Nor is it surprising that the soldiers were lawless, seeing that they were pressed for service, clothed and sent out without any training or drill; nor were their officers educated in any way to organise or command large bodies of men.

1601.

The preceding eight reigns had been fairly peaceful. It is said unless foreign wars occur at intervals, to exercise the military ardour of the people, civil war will inevitably result. The absolute despotism of the Tudors had given place to the milder government of the Stuart dynasty. None of the Stuarts were successful as Military leaders, they lacked the firmness and self-confidence necessary for such duties.

1603. When the Scotch King James VI. became James I. of England, he does not seem to have troubled about military affairs ; his coronation was at Westminster Abbey on 25th July. In his speech to the House of Lords, he speaks of the horrors of civil war, the blessings of peace, and the benefit which would come from the union of England and Scotland. Articles of Peace with Spain followed the next year, but of Military matters we have nothing until 1614.

1614. A general Muster of horse and foot all over England was commanded by the King, after harvest, in which every county took part. In many places they had to buy new weapons and clothing ; most had decayed and some had gone out of fashion.

The London soldiers were considered so excellent that many country gentlemen went to see them exercise in the Artillery Garden, without Bishopsgate (which had been used for practising artillery from the year 1586). They returned to their own counties to model the Militia on the same lines as these London Trained Bands.

From this period, until Charles I. became monarch, history gives no further information as to soldiers or musters. It is only reasonable to assume that the Trained Bands of England met at certain periods, especially in those counties which were able to pay for their maintenance——in Berkshire we know they exercised annually—for when Civil War broke out, both King and Parliament were able

c

to raise troops, which could not have been done unless the Militias had been kept together to some extent previously.

1624. The old town armour of Reading was sold and new bought.

Even at this date the Musters were held in the Forbury in the springtime. In some evidence in a law suit, given before the Mayor, the witness mentions "the Muster in the Forbury," which must have taken place the middle of June that year.

17th and 22nd November. A warrant in Reading was issued to impress fourteen strong and able men fit for the wars, to serve under Count Mansfield, and 40s. of money, being 5s. each man, for eight of them. All charges were to be paid until they were delivered into the hands of their Captain. This was done under the Lord Lieutenant's warrant to the Mayor, and the inhabitants had to pay; the corselets cost 18d. each man, and the muskets 12d. per man. Count Mansfield, or Mansfeldt, was a hero of the Palatine war; he was employed raising troops on the Continent to expel the Spanish from the Netherlands. He embarked in the autumn from Zealand to collect the English money and troops promised him, but was wrecked; his English captain and crew were drowned, he and a few followers escaped in the long-boat, and landed in England. He was promised £20,000 per month, and 12,000 soldiers, by King James, and these men were levied by press. Untrained and undisciplined they were sent to Dover (where several were hanged for misdemeanors and mutiny) thence the army embarked. A landing at Calais was refused them. From thence they were taken to the Island of Zealand, where the Dutch were as little ready to welcome them as the French had been; sickness broke out, for the transports were miserable and ill-adapted for large bodies of soldiers, who were crowded together between decks. At last the expedition reached the Rhine, and the border of the Palatinate. By this time one half of the army had perished from disease, and warfare was impossible. Part of the expense of the Reading contingent was paid by the town, and part by the county. Alas! very few of the men thus sent lived to return home, and those who did must have had terrible tales to tell of the privations and horrors they had endured.

December 16th. Two hundred soldiers had been raised in Berkshire, and were commanded by Captain Francis Bassett. The King sent his Council of War a list of their names and the places they had come from in Berkshire.

The names of the men impressed, December, 1624, were : For the Town—Hugh Sherwood, Thomas Johnson, Justman Edmondes, William Webb, Edward Hudson, Andrew Browne, John Jerome, John Cowbery, William Berry, Nicholas Addams, Thomas Taylour, Thomas Wigmore, alias Ingleton for Tilehurst, William Collett, Patrick Hackett, John Hoodd, John Dolman, William Bewell, John Mullyns, Sacarye (Zacharias) Max, John Bardye, George Wylmore.

For the Country — George Willyamson, John Hatt, Anthony Belgrove (Blagrave?), John Perse, John Woodare, John Moyle, Roger Tubbe, Thomas Wirge, Edward Bulley. But the country did not pay towards keeping the eighteen men the thirteen days. The total expense came to £29, of which £26 came from the Town, 40s. from the Hall cofferer, of which was paid to the receiver for coat and conduct money, £22.

By taking them to Newbury in June, 1625, it looks as if the Muster took place there that year.

The Warrant for the Muster of the Trained Band, was issued in Reading, Monday, June 27th, and ten 1625. men for supplies appointed, and the four to be trained for the Hall also appointed. The plague was bad in Reading, and every precaution was taken to prevent it

C2

spreading.   This may account for the soldiers being sent to Newbury.

1625.  May 7th.   Eighteen men of Reading were raised for service.   Two years later thirty more were sent. The Military history of Berkshire is only the history of England in another form; for every war, men and money had to be found.   There was no regular army, the trained bands being the Militia under another name.   The safety of England as well as her honour and glory depended upon the loyalty and good conduct of undisciplined soldiers. The eighteen soldiers above mentioned in 1625, were only the Reading contingent, the whole regiment of Berkshire men numbered 200 under Captain Reade Wildgos, they were ordered to proceed at once to Plymouth.

May 29th.   William Lord Wallingford was at that time Lord Lieutenant of the County.   Shortly before, Berkshire had been dismayed by a demand for 3,000 armed men for the King's service, but there were only 1,000 soldiers in the whole county, and those the trained bands, there was no money to pay a large body of men, and even if the money had been forthcoming, it was impossible to provide arms. Further it was said, if men were to be levied and sent out of their own county they ought to be given a month's pay in advance.

June.   The Deputy Lieutenant, Sir Francis Knolles, and Sir Richard Lovelace sent a warrant for the impressment of eighteen strong and able men in Reading fit for service, (but none of them to be of the Trained Band), 40s. per man for eleven of them was to cover the cost of coats and conduct money, etc.   They were kept in counter thirteen days, and were then sent to Newbury.

September 9th.   The constable of Reading sent in his bill for impressing ten men.   These also went to Newbury. Their coats cost 16s. each.

The men were Isaac Croome, John Nicholls, John Belson, Edmund Daling, Anthony Porter, Richard Poole, Thomas Powell, Christopher Pryer, Roger Camyll for Benham (Beenham), Henry Homan for Whitley.

January 30. By direction of the Deputy Lieutenants, Sir Francis Knollis and Sir Thomas Vachell, the Mayor of Reading reviewed the arms of the Town, and warned all defects so that they might be ready at an hour's notice.

1626. March. The soldiers pressed in Reading were William Littlepage, William Pearse, Thomas Creed, William Collett, Luke Payne, William Boone.

August. The return of all able men between sixteen and sixty in Reading, made by the Constables to the Deputy Lieutenants, amounted to 1,080 men.

October. A tax was made among the inhabitants of Reading to pay for the lodging of thirty-seven soldiers at 8d. each, and their horses and carriages, and all other charges. Troops on the march were paid for by the Corporations of the Towns at which they halted *en route*, the expenditure was paid by a tax levied on the inhabitants. These taxes were difficult to raise, being looked upon in the light of war taxes levied in times of peace.

A Warrant from Deputy Lieutenants of Berkshire in November. Captain Gifford and his company of eighty-eight soldiers billeted in the Town of Reading. Every soldier and serjeant was allowed 3s. a week, the Ensign 5s., the Lieutenant 7s., and for the Captain and other superior officers as they agreed for themselves. Estimated to cost £7 a week, which had to be repaid by a tax levied in the town.

These soldiers went from Reading towards Winchester, April, 1627.

1627. May. Soldiers were again furnished by the Town of Reading. Six soldiers by name—Edward Cooper, John Lidyard, William Polman, John Assone,

Thomas Clifford, Robert Cooper. Fifteen others pressed were for the Town and Country.

Out of five soldiers, three took the press money and then ran away. The men had to be at Newbury by September, 1627.

In October of the same year, ten able men had to go to Newbury by 20th of the month, at the cost of £9. £6 payable by the Town, and £3 by the Country. Henry Ingane, a tapster; Richard Gayger, a weaver; John Webster, shoemaker; John Kenton, weaver; Peter Harvey, tooth drawer; Thomas Powell, a vagrant, were thus impressed; and for the Country—Ebson, Coanes, and Carter.

March 7th. There was a controversy between the young men of the town, led by two apprentices, Joseph Fillett and John Richards, against the soldiers billeted in the town. Apparently the apprentices were in the wrong, as they had to find sureties for good behaviour, and not being able to do so were committed to the counter, in other words, sent to prison.

Two young men, John Barker, junior, and William Booth interfered with the soldiers who were playing football in the Forbury, and tried to take their ball. It caused much trouble to the constables and officers, and danger of hurt to many others, for naturally the soldiers resented such uncalled for interference, and a general free fight ensued.

1628.    Warrants from Deputy Lieutenant for billeting and lodging ninety men, part of Colonel Ramsey's Company in Reading, at 3s. 6d. per man every week. Captain Roger Powell was Commander of these ninety men, no doubt on the march to join his Regiment.

The billeting of soldiers in private houses was one of the subjects of offence between Charles I. and the English people. They dreaded the soldiers, who were under very little command, for this reason the establishment of a standing army was regarded as dangerous to public peace.

Ship money—that hated war tax levied in a time of peace—was used to pay for the soldiers. Its unpopularity was greatly due to this dread of a permanent army, which also was looked upon as a slight on the protective powers of the old regiments, who in former times had done good service for the Crown.

June 9th. The shopkeepers in Reading were privately warned not to sell any match or powder to any soldiers in the town, or any other person for the soldiers' use. Every man in the Town had to provide match, powder, and shot ready in his house.

December 9th. Warrant from the Deputy Lieutenants of Berkshire. Colonel Ramsey's Troop of ninety men, part of the Troop billeted in the Borough from 22nd January to 26th July, 1628. Twenty-six weeks and four days at £3 10s. per week.

Soldiers were sent from the West of England to Kent, Surrey, Sussex, Bucks, and other Eastern parts, passing through Reading. These were paid for partly by the town and partly by the county.

Forty shillings was collected in July to pay for the Beacons at Cutchinsloe being watched. Other Beacons there must have been in the county, but none other is named.

The third Parliament called by Charles I. was only held 17th March of this year. Previous to its assembling, the King had levied his War Tax. Besides having a large number of soldiers at his command, he suggested the probability of bringing over German troops of horse to England. Martial Law had been introduced, the King and his people were on bad terms ; Parliament was determined to assert its privileges, while the King was equally determined to yield none of his. For eleven years he would have no Parliament, and during this time he levied taxes at his own discretion, and the discontent of his people was growing more and more visible.

**1631.** Six thousand English Volunteers were levied for service of the King of Sweden; among these Berkshire sent its share.

June. Ten or eleven voluntary soldiers were apparelled for Captain Hammond's Company, of Colonel Hamilton's Regiment in Reading.

Colonel Robert Hamilton was, in 1647, Governor of the Isle of Wight, when Charles I. was taken a prisoner there. He was Military Governor of Reading, 1648, High Steward and Member for the Town, 1654 ; he died that same year.

**1633.** That the King was not unmindful of the Army and its improvements is certain, from the fact that he issued an order that the time and step of the marching of regiments should be the same, both for regiments at home as well as those on foreign service. Negligence and want of time on the part of the drummers had caused much confusion, which was felt directly troops were put to work together. This order is the only one I have found regarding marching, or relating in any way to the bands of any regiments.

The old armour in Reading was not to be sold, but repaired and kept clean. In the unsettled state of politics, it was necessary to be ready.

**1637.** A petition by Sir Edward Clerke to the Justices at Sessions, shewed that for eighteen years past the Town had paid 8s. rent for housing a parcel of gunpowder for the County, also, in another room, matches, mattocks, shovels, and other things, worth then 18s. 4d., now 20s. per annum. Thus in that space of time there was a decided rise in house rent in the Borough of Reading.

The County Magazine is mentioned, 1637, a barrel of gunpowder being sold to the Town of Reading from it. Each troop though raised in the county, was trained in the town of their division. Even as late as the 19th Century, the Berkshire Militia was divided into companies, each known by the name of the division in which it was raised.

# CHAPTER III.

## THE CIVIL WAR.—1640-1649.

IN 1640, the Abingdon Division had to find 240 men for the Army, but as neither coat nor conduct money was properly collected, it is not surprising that the force was in a bad way, and only about 120 men who came from the neighbourhood of Radley came forward. The Reading Deputy Lieutenants promised to send their men to Abingdon (then the County Town), as soon as possible, but as they named no day, the promise was a vague one, dependent upon how soon, if at all, they could gather in sufficient money to cover the expenses.

Captain Belloes (Bellasis) was sent to Reading to hurry them, and brought back word that a hundred men were ready, and the rest would soon follow.

In the Forest division the state of Military affairs was bad, Captain Andrews who commanded that part was a recusant. Moreover, he had had a dispute with one of his soldiers named Bates, and an inquiry had been held on the subject. Neither Captain nor Lieutenant appeared to defend themselves, Bates had been imprisoned for a week, and the Captain had evidently been much to blame. Only £120 out of £300 had been raised; the constables returns were imperfect and neglected.

In Oxfordshire, things were even worse, for 116 men were forced into the King's service, otherwise they would have been dismissed again to their homes, so with great difficulty the authorities were obliged to get coats made for them. No mention is made of any other item or garment, evidently the outfit was extremely scanty, and probably consisted

only of the coats, which the authorities lay so much stress
upon.  Perhaps the officers of the troop had to provide the
men with clothing, for certainly no mention of any is made
in the accounts, and as the men were so poor they could
not have clothed themselves.  However, by July 1st, the
Regiment was officered, and thus gathered together was
sent off at once, without drill or discipline, under Colonel
Sir Jacob Astley, to join the Royal Army on its march
northwards to repulse the Scotch Covenanters under Leslie,
who had crossed the border and were marching with all
speed towards York, the northern capital of England.  The
Scotch Army was well drilled, a contrast to the English
troops ;  they were well organised, and though the English
Army numerically was far superior, yet as we shall presently
see, it was stronger on paper than in reality.  The Lord
Lieutenant of Berkshire, the Earl of Holland, had command of the Horse in this expedition.

The Berkshire men advanced with all speed northwards,
but all along the route they met with discouragement from
the country people where they halted.  Ill clothed, ill paid,
and doubtless ill fed, they resented being taken against their
will away from their own county, to fight what almost
amounted to a civil war, and by many was regarded as
wrong from a religious point of view.

July 13th.  When they reached Northamptonshire the
spirit of discontent and unwillingness took definite shape.
At Brackley, and at Daventry, the men of the Reading
division, under Colonel Sir Jacob Astley, refused to be
taken any further, saying they " would not fight against the
Gospel," and knew they " were going to be shipped and
commanded by Papists."  Neither threats nor persuasions
availed : when a Berkshire man does make up his mind, he
never changes it.  The spirit of revolt was rife among the
whole army.  Finally the men of Berkshire and Oxfordshire left their officers and, disbanding themselves, set off

homewards. It is said that afterwards they were sorry for this act of insubordination, and would fain have returned, but they feared the severity with which they would have been punished.

Martial Law was proclaimed in the Army. Seven of the ringleaders were captured, imprisoned, and executed as examples. Sir Francis Knolles called a special meeting of the Deputy Lieutenants of Berkshire and other places, to consider the situation, and they decided to arrest the men as soon as they returned to their homes, but this the men had expected, for it was said they all went towards Somersetshire and the West of England, instead of returning to their own counties.

Thus the King found himself minus about 1,000 men, his army weakened, and the discontentment spreading. He issued fresh orders for more Militia men to be raised, and Lord Holland, the Lord Lieutenant of Berkshire, again endeavoured to execute the commission. At heart, he probably was on the Parliamentary side, for some two years later, soon after the Civil War broke out, he left the King's party and sided with the Commons. A troop was raised, however, though the Deputy Lieutenants had great difficulty, especially with the Vale men who were less ready to come forward than the other divisions of the county. By this time the rupture between the King and the Parliament had assumed a serious aspect. Parliament had again been summoned after a lapse of eleven years. Politics ran high, *Cavaliers* and *Roundheads* became recognised party names. The King refused to yield any thing of his power over the Army. When a definite answer was demanded by Parliament, as to the right of calling out the Militia, he deferred his answer; no remonstrances were of any use, and the situation grew worse and worse.

1642. Preparations for Civil War began long before the actual fighting. The King and Parliament were both arming themselves. In Berkshire, the Earl of Holland raised the Militia for Parliament. Probably some of those very men joined him who had seceded from the Royal Army at Daventry in 1640.

It is said the real point of contention between King Charles and his Parliament was the right of calling out, and having full power over, the Militia. Any way, it was the Militia Troops of England who fought the war and settled it, for there was no standing army of any strength at that time.

The actual outbreak of the Civil War began in July, 1642. Parliament sent a Commission to King Charles at Beverley entreating him to forbear his hostile preparations. He retorted by telling the Commissioners that Parliament ought first to submit to him. Neither side would give way, thus war was declared; fighting began at Hull, where the Governor closed the gates and refused to admit the King. At Nottingham the Royal Standard was unfurled on 25th August, and two months later the battle of Edgehill was fought between the rival armies. Sir Jacob Astley's prayer, as he advanced to the battle was short, fervent, and to the point, " Oh Lord ! thou knowest how busy I must be this day. If I forget thee, do not thou forget me. March on boys ! "

Winter was fast approaching, by this time each town and county had declared for one or the other side. Reading belonged to the Royalists, and was garrisoned by them, although the townspeople, with the exception of the Mayor and Magistrates, sided with Parliament, but had not openly declared.

Failing to get near London, Charles made his winter quarters and garrison at Oxford.

Sir Jacob Astley still remained with the Royal troops, and was made Major General under the Earl of Lindsey.

Apparently the Militia was entirely reorganized by the Parliamentary Militia Commissioners, as Major Evelyn seems to have had sole command, and no other Colonel appointed to it after Sir Jacob Astley.

During the Civil War the troops were kept as garrisons for the different towns, and the battles or skirmishes were of the nature of sorties from the garrisons.

Berkshire was much divided in politics during the Civil War. On both sides prominent leaders and officers were natives of the county. The Earl of Holland, the Earl of Berkshire, Sir Robert Pye, Daniel Blagrave, Edward Bayntun, and many others, figure in the various accounts of the time.

Wednesday, February 15th. About this day a warrant, directed to the Mayor of Reading, from the High Sheriff of the County of Berks, touching the speedy raising and advance of the Berkshire Regiment for his Majesty's service and defence of this County was openly read, and then it was agreed that Mr. Mayor should make a return according to the meaning of the said warrant.

June. The King and Prince Rupert sent requiring contributions from Reading and the neighbouring hundreds to pay for the fortifications and garrison at Wallingford.

The Reading division of the county had to find 500 men for the King's service, and twenty-five of that number were ordered to be at Abingdon immediately.

July. Charles I. made the Marquess of Hertford, Lord Lieutenant General of the Southern Counties and seven Counties in Wales.

July 12th. Parliament voted the raising of an army, professedly, for the safety of the King's person and the defence of the country and Parliament. Essex was created Commander-in-Chief, under the title of Captain-General ; to him was entrusted, with the aid of committees, the nominations of Colonels, Field-Officers, and Captains, to serve under him. Many members of Parliament volunteered their services as soldiers. The King, meanwhile, issued a

"Commission of Array," and moved about the country collecting men and levying taxes for the equipment of his army.

October 1st. O'Neal, Sergeant-Major to Count Robert, better known as Prince Rupert, sent a letter to Mr. Vachell, the High Sheriff of Berkshire, commanding him in the King's name to raise the power of the country (in other words, the Militia) to conduct the King through it, but the Sheriff stayed the messenger, refusing to obey the order. Hampden, one of the principal leaders of the Parliamentary party, married an aunt of Mr. Vachell.

In October a skirmish took place at Kingston in Surrey. The Trained Bands of Berkshire, as raised by Lord Holland, and those of Surrey formed part of the Earl of Essex's army against Prince Rupert; the former were victorious, although they lost about 300 men in the fight.

November 1st. A party of horse from Abingdon advanced to Reading. The Parliamentary Governor, Henry Martin, evacuated the garrison without orders from head-quarters, an act due, it is said, to cowardice on his part. Thus for awhile Reading was again in Royalist hands.

November 4th. King Charles himself came to the town on his way to London, and again at the end of the month, when he made Sir Arthur Aston Governor, with a garrison of 2,000 foot and a troop of horse.

The old difficulty of clothing the soldiers still gave great trouble. On November 8th, the King issued a warrant to impress tailors in Reading and within six miles of the town, to make clothes for the soldiers; the constables of the district had the unwelcome task of seeing that the order was enforced.

The Mayor and Aldermen were also ordered to seize as many carts or boats as might be necessary, so as to send the suits of apparel as quickly as possible for the soldiers, and for the transport of provisions for the garrisons.

This warrant, no doubt, caused immense discontent, for later on the clothiers were promised that no more clothes should be seized without payment being made for them, if they in return promised not to send any of their goods to London.

November 7th. The King ordered that the inhabitants of Reading should bring all their armour and arms to the Town Hall by one o'clock. Any house in which arms were found after that hour was to be given up to the soldiers to plunder. •

Another sortie from the town of Reading took place in February. Some 500 Dragoons and three troops of horse met the Parliamentary forces in Henley, when the Royalists were defeated and lost two officers and several men. About a week later Sir Jacob Astley, the former Militia Colonel, with a party of men made a raid out of the Reading garrison as far as Old Windsor, taking away all the horses and cattle they could find; which they did without the garrison at Windsor Castle knowing anything of their visit till after it was over.

Taxes were levied in Reading for the support of the different garrisons in the county, Wallingford, etc., but the town pleaded too great poverty to pay them.

After the battle of Alresford, Lord Hopton brought his troop back to Reading, from thence to Abingdon; they had originally been drawn from Berkshire by levies.

November 10th. There was really a chance of cessation of war. The Commissioners of Parliament advanced to meet the King with conditions of peace; an interview took place at Colnbrook, but the King delayed an immediate answer, and instead of waiting for further conference, advanced as quickly as possible towards London. The two forces met at Brentford and a skirmish ensued; and the following day at Kingston the two armies met face to face. Indecision on both sides saved a terrible battle, but only to prolong the war.  The King withdrew to Hampton Court, where, after spending two nights, he retired to Oxford, *via* Reading.

December.  Hampden had hurried towards Reading, hoping to intercept the Royal forces, but Prince Rupert fearing this, left his baggage in Reading garrison with a guard under Colonel Lewis Kirke.

As Hampden's connections belonged to Reading, and Kirke was a stranger, the chances were in favour of the former, but a determined stand was made to retain the garrison for the King.

After four hours battle the Parliamentary army again took possession of the fortress.  Some 400 men were slain, and Colonel Kirke fled with such remains of his garrison as had not been taken prisoners, to join the King at Oxford.

The account of this attack is to be found in the *Life of John Hampden*, but it is evident that the Parliamentary party must have lost the town again, for in the following April the Earl of Essex appeared 1643. before Reading.  The garrison then contained over 3,000 foot and 300 horse soldiers.  The Governor, Sir Arthur Aston, refused to surrender, although they were short of provisions and ammunition, and the attacking force was more than double in number, the whole of the Earl of Essex's army being brought thither.  After a siege of four days the garrison surrendered ; they were allowed to retire to Oxford with colours flying and bands playing.  The Governor, Sir Arthur Aston, having been badly wounded in the head by a falling tile, and so prevented from active duty by concussion of the brain resulting from the blow, the command of the garrison devolved on Col. Fielding, the oldest officer there, who, taking fright at hearing of Essex's advance, hung out a white flag and agreed to capitulate. Reading then became a Parliamentary stronghold until after the first battle of Newbury, fought September 20th.

September 20th. The first battle of Newbury took place. Of the Militia who fought there one can only judge by the names of the commanding officers, namely: Sir Jacob Astley, Major General; Lord Hopton's Brigade, to which, as we know, men had been sent from the garrison at Reading; Colonel Bellasis' Regiment; Colonel Bowles' Regiment, both these Officers had been stationed in Reading. All these names have been mentioned in connection with Reading soldiers and its garrison. Captain Robert Hammond was also a Governor of Reading later on.

Both sections of the Berkshire Militia fought in the battle, that raised by the King, as well as the Parliamentary division. The Earl of Holland left the latter and took the side of the Royalists shortly before the battle.

Though this battle was an important one, it was not decisive; the King's losses were severe, but he was left in possession of the town of Newbury and the battle-field, whilst Essex marched along the Kennet Valley on his way towards London. It may have been then the skirmish took place at Theale, in the lane still called "Dead Man's Lane," Prince Rupert attacking his rear guard, slew many and took others prisoner. Essex reached Reading the same evening, but only remained there two days, continuing his march to London. Whereupon Reading was regarrisoned by the King's forces, some 3,000 men and 500 horsemen, under Sir Jacob Astley. Another authority gives the number as 2,500 men.

Two days after this date John Hampden died from the wounds received in the battle of Chalgrove. The news of his death was received with universal sorrow. His regiment —called the "Green Coats," no doubt from their uniform— was one of the best in the Parliamentary army. The tide of war again set towards the North of England, where Cromwell was with his soldiers. In the West, Prince Rupert took Bristol, and, at this time, it seemed as if the Royalist party would carry all before them and gain complete victory.

D

April. The garrison was in the Forbury or
1644. Abbey ruins. Coals and firing were provided
by the Corporation for the Guard, which nightly kept watch
all round the town to protect it from surprise.

Part of the garrison joined Lord Hopton, so the King, by
letter from Oxford, requested a certain number of Towns-
men to keep watches, and a levy of men to be made to
replace those who had gone with the army, as Reading was
an important stronghold, lying as it does midway between
London and other points of vantage.

April 14th. A regiment of auxiliaries were ordered to be
raised by well-affected persons in the town and neighbourhood of Reading, under the command of Colonel Richard
Neville, the High Sheriff, to be kept entirely to garrison the
town. Sir Jacob Astley was ordered to return there at once
with arms and ammunition, and they were to provide
muskets for the men at 12s. each.

In May, the King, finding his supplies failing, demolished
the fortifications of Reading, evacuated the town, and
marched to Oxford.

Once more Essex sent and took possession of it, he being
then at Windsor. For the rest of the war Reading remained
a Parliamentary stronghold.

June 26th. The contributions for the garrison at
Wallingford, due from Reading, were not forthcoming, so a
party of horse came—June 3rd—and carried away the
Mayor of Reading, William Brackstone, a prisoner to
Wallingford until the contributions of £50 weekly, due from
June 3rd, were paid up. The Alderman wrote their utter
inability to pay the money and this letter was Lieut.-Col.
Lower's reply:

"Their letter gives him little satisfaction since they do not
so much as promise their endeavours to raise what is required.
If they will send £100 he will forbear the rest and abate the
weekly sum of £50, till then the Mayor will be detained. If

we shall hear you contribute anything underhand to the Rebells, we shall require it double, but I have a better opinion of your loyaltie and affection to the King's service."

The King's methods of raising money were decidedly high-handed. The fact was, he was so impoverished that with difficulty the troops were kept together; clothing, food, and raiment being scarce sufficient for the needs of the army.

It is said that each company of a regiment carried a colour until the 18th Century. The Berkshire Militia were divided into companies, called after the division from which they were drawn. The standards were most likely the heraldic devices of each town. At this time the Regiment was so split up that its history is well nigh impossible to trace. There were two distinct "Berkshire" Regiments. On the King's side, probably little reliance was placed on it, and the men were, perhaps, drafted into other regiments or commanded by strangers; the old Regiment was with Essex's army.

July 26th. Parliament ordered a general Muster to be held in Reading, at three o'clock that afternoon, of the Trained Bands; fighting having again begun in the Royal County.

Basing House and Donnington Castle held out obstinately for the King, who seemed still to have some small chance of success, a hope crushed after the battle which was fought at Newbury in October, whence the King had to fly to Oxford for safety.

1645. June 3rd. Proceedings at the Committee of both Kingdoms. To write to all the commanders of the Parliament's garrisons in Counties Oxon, Berks, and Bucks, to obey such orders as they shall receive from Major General Browne.

Major-General Browne being by ordinance of Parliament appointed to command in chief the forces within the three

Counties, Oxon, Berks, and Bucks, to the end that he may the better carry on the service in those parts, we desire you to receive and obey his orders. Sent by Major-General Browne.

June 11th. This day Captain Goddard brought word from the Governor of Reading, that he would not protect any soldier from being arrested but those that were in his own company, or that did receive pay from the Committees ; thereupon Mr. Thomas Harrison proceeded in his action against John Webb. This, probably, refers to the petition sent by the Reading Corporation to Parliament, December 2nd, 1644, for the relief of the town against the insolencies and violence of the soldiers.

A tax, ordered for the relief of the British Army in Ireland, was levied in Reading.

Charles I., Vol. DXI., No. 29. October 25th. The Committee of both Kingdoms to Colonel Martyn :

> "We wrote to you yesterday to march with your whole regiment to Donnington Castle, and these are only to signify to you that the rendezvous of all the forces of the three counties, Oxon, Berks and Bucks, that are appointed for that service is to be at Reading upon 28th present, where we desire you not to fail to be, lest that service of so great concernment be retarded."

Writ against Capt. Curtis and James Maynard, to enforce payment of £50 which they owe the Corporation of Reading. The former was probably an officer of the garrison who lived in a private house in Reading, rented from the Corporation.

November 29th. At this time was produced before the Corporation of Reading a writing signed by Colonel Fortescue, in these words, viz. :

> "You are to march five companies of my Regiment to Reading and to quarter there till further order.

> "RICHARD FORTESCUE.

"28th November, 1644.

"To Captain Leverington or the chief officers present."

It is possible this name ought to be Levingston, and that he was of the same family who inherited Tidmarsh later on from Sir Peter Valore.

The Berkshire detachment of Horse was 300 strong. It was under the command of Captain John Blagrave, of Reading, who was promoted to the rank of Major. They fought for Parliament in the second Battle of Newbury and at the Siege of Donnington Castle.

1646. Money was raised in Reading for the relief of the British Army in Ireland ; again this tax had to be paid.

July 8th. The Mayor of Reading determined to attend Sir Thomas Fairfax about the aspersions laid on the town, and about freeing it from the Military.

Captain Robert Aldridge complains through his Cornet, William Ivery, that his horse had been seized as a deodand by mistake ; and this complaint, being just, was corrected.

The Corporation were anxious to get rid of free quartering of soldiers in Reading. The plague had broken out in the town. So many soldiers had been sent there that they could not find quarters in the inns, and the old grievance of billeting in private houses had never been given up, at which the people murmured.

Captain Morris required allowance for free quartering in the borough of three sergeants, two drummers, three corporals, one gentleman of pikes, one gentleman of arms, three-score common soldiers.

The sergeants were allowed 5s. 6d. per week. The others 4s. per week, and the 60 soldiers each 2s. 8d. per week. Total, £10 4s. 6d.

Upon Captain Morris being asked by what order or authority he required this, he said he had no order under any man's hand, but would bring an order some time the next week, whereby to enable the Mayor and Aldermen to tax and levy this money upon the inhabitants ; and for the

Company's discharge, £5 was advanced to Captain Morris for immediate necessaries. Captain Morris brought the warrant as he promised from Thomas Barham, Quarter-master-General to Major-General Browne, for quartering Colonel Baxter's soldiers in the town, whereon it was agreed to pay £10 4s. 6d. for three weeks. At the end of the three weeks the tax had not been gathered to pay this sum, so the Corporation resolved that the money already paid and the names of those who had not paid should be delivered to Ensign Larkin, and that he or some other officer should collect this rate and pay the soldiers, or else place soldiers in the houses of those who refused to pay. If the money was not forthcoming the Mayor was to pay it, and then repay himself out of some other tax. At last a 2d. rate was levied for the purpose. The Colonel's name is spelt Bacster, Baxter, or Barkstead. In 1651 his name is spelt Barkstead, for a letter was sent to him relative to some business connected with the lease of John à Larder's lands, the tenant, Mr. Thomas Harrison, having died; he also expressed his readiness to promote the business of the Corporation as regards Kendrick's Charity, being at that time Military Governor of the town.

Colonel Barkstead was one of the eleven Major-Generals appointed by Cromwell after the riots of 1656, and finally had to fly from England at the Restoration.

A dispute arose in Reading about coals having been taken for the soldier's use by some of the King's officers, but the charge was dismissed on enquiry into it.

The plague in Reading again was bad, and caused much anxiety to the Corporation.

There was a meeting of churchwardens and overseers from each parish in Reading to decide how many soldiers were to be quartered in each parish, and lists were drawn up of each ward, with the accommodation it afforded for billeting.

Reading was further taxed for the disbanding of the garrisons of Abingdon and Reading, the sum of £170 13s. 4d.

The petition of William Hill, Symon Costen and Michael Holman, being of the Sub-committee of Accounts for County Middlesex:

> " That by ordinance of Parliament, 18th July, the Committee of Middlesex without the line was charged with 20 light horse and 12 dragoons, towards the keeping of the garrisons in Counties Oxford, Bucks, and Berks. That the Committee have charged each of us towards this number one light horse or £12 in lieu thereof, represent that this charge is unreasonable and disproportionate."

October 9th. A tax was levied for quartering Captain Morris's company, consisting of three sergeants, two drummers, three corporals, one gentleman of pikes, one gentleman of arms, and 60 common soldiers in Reading. They had been three months in the town.

1647. August 30th. It was agreed by the Town Council of Reading that a petition should be forthwith presented to the Hon. Sir Thomas Fairfax, for the vindication of the Company and town from the aspersions laid upon them by Captain Goddard and others, and that the Company or the greater part of them should go to Sir Thomas Fairfax. This "aspersion," whatever it was, had been troubling the Corporation of Reading for a long time.

August 14th. Copy of an order from Sir Thomas Fairfax to Colonel Vincent Goddard, to take account of the arms in the custody of the Mayor of Reading. On August 18th, Mr. Curtis went into the Armour House under the Hall, but found no arms there.

The King was now completely in the hands of his enemies. His imprisonment began in the Isle of Wight, at Carisbrook Castle, where Hammond was Governor. He did not despair even then, as various attempts were made to rescue him from his adversaries, but these were without avail.

1648.     May 8th, Windsor Castle.   Letter from Sir T.
Fairfax to Colonel Harrison, " or the chief officer
with his Regiment at Gilford " :          •

> " I have received information this morning, from some of
> the well-affected in Reading, that there hath been of late several
> meetings of cavaliers and other affected persons in a hostile
> manner in that town, and that they have threatened to make the
> same a garrison for the King.   I desire you would, upon receipt
> hereof, send two troops of your Regiment into Reading, and to
> quarter them until the Major (Mayor) and well-affected may put
> themselves into a posture of defence to preserve that town for
> the service of the Parliament."

This order was to be communicated to the Mayor of
Reading, who was to be assisted in apprehending such
persons as might disturb the peace.

August 21st.   Domestic State Papers, Charles I.   Vol.
DXVI.   Proceedings of the Committee of both Houses
at Derby House :

> " No. 2.   That Quartermaster-General Fincher do go down
> into Berks to-morrow after the Houses are risen, taken with him
> 20 horse of his own troop; and that letters be written to Counties
> Surrey, Wilts, Hants, Berks, Oxon, Bucks and (the garrison at)
> Wallingford for forces to assist him, and he is to carry them with
> him.

> " The Committee of both Houses to the Committee of
> Surrey, Wilts, Oxon, Berks, and Bucks.   The Commons having
> notice that there are some forces risen in Berks, under Colonel
> Marten Eyre (or Ayre) and others which commit great outrages,
> have appointed the horse of several counties, including yours,
> for the suppression of them, and directed Quartermaster-General
> Fincher to be sent down to command the party appointed for
> this service.

> " You are therefore to repair to such rendezvous as he shall
> appoint, and receive and execute his orders in the prosecution
> thereof."

About this time the Earl of Holland transferred his allegiance from Parliament, whom he had assisted all through the war, and he joined the King's side, only to be made a prisoner like his Royal master ; like whom, also, he suffered death on the scaffold.

1649. The trial of the King was begun early in the year. Its tragic ending finished with the King's execution on January 30th, after which England was declared a Commonwealth, with Oliver Cromwell as its head with the title of Protector.

The Royalists evidently never repaid the Corporation of Reading for money spent on the soldiers, nor could any redress be obtained by going to London, for Parliament did not hold themselves responsible for the King's debts.

The Corporation advanced money, nevertheless, to a troop of horse soldiers in the town (September 14th, 1649, and £100 to the foot soldiers on November 5th), to avoid free quartering, and of course the inhabitants had to pay the same back in inevitable taxes.

November 14th. It was agreed that all troopers and foot soldiers then quartered in the town, which had been there seven nights already, should be forthwith billeted in the inns, alehouses and taverns of the borough, except such private houses as they can agree with.

The war was at last over, and Reading settling down again in peace, to recover the terrible effects of Civil War.

# CHAPTER IV.

## UNDER THE COMMONWEALTH.—1650-1659.

THIS day, February 5th, 1650, the Mayor of Reading (Mr. James Arnold), having received a warrant from the Commissioners of the Militia of this County of Berks for the assessing and collecting of £56 5s. in the borough, for the furnishing drums, colours, trophies, ammunition, and other emergencies for the safety of this county, did sign a warrant for this assessment.

No sooner was the Commonwealth actually established, than they set to work to bring about those measures which they had tried to force upon the King for many years. The Militias were placed under "Commissioners."

In the past years of fighting and confusion, there appears to have been two separate Militia corps in Berks, one raised on the Parliamentary side—the other recruited for the King. That re-organization took place in 1650, there is ample proof in the appointment of officers made that year.

The alteration in the Government of the country changed everything, just as the Mayor and Corporation had to replace the Royal Arms by those of the Commonwealth, so the Militia had to march under new standards, clothed in new uniform ; no longer governed by the Lord Lieutenant of the County, but by a Parliamentary "Council." The cavalier soldiers are always represented as wearing armour, and differing in many points from the Roundheads. After this time, we hear no more about the town "armour" of Reading.

*From a portrait at Aswarby Hall.*

COLONEL WHICHCOTE

STATE PAPERS, DOMESTIC.—1650-1660.

August 23rd. Militia Commissions granted by Council of State :—"Berkshire : Troop H, Captain Arthur Evelyn, Lieutenant William Stephenson, Cornet Richard Greene ; Troop F, Captain Vincent Goddard, Lieutenant Andrew Keepe, Ensign Arthur Horne. October 30th, Colonel Arthur Evelyn, Lieut.-Colonel John Blagrave, Major Vincent Goddard."

The re-organisation under the Commonwealth was, perhaps, as far as the officers were concerned, merely a matter of re-appointment, with additional commissions granted ; Major Evelyn had been connected with the Regiment many years, as had also Captain Vincent Goddard. All the names above-mentioned sound local. Major Evelyn was raised to be Colonel. The Blagraves were rich clothiers of the town, and so too were the Goddards. The other officers I have not been able to identify.

Vol. IX. Council of State, Day's Proceedings, April 4th. No. 6 : " That the names now given in by the gentlemen of Berkshire to be commissioned for the new Militia in that county be approved of."

Note the term, " New Militia," above. The name, Militia, is supposed only to date from the time of Charles II. ; but all Encyclopœdias are most vague and incorrect on the subject.

August 23rd. No. 7 : " The names sent in by Major-General Harrison for a troop of horse, and a company of foot for Berkshire, approved and commissions issued."

Vol. IX. Council of State. Day's Proceedings, October 1st. No. 6 : " Colonel Christopher Whichot (or Whichcott), Governor of Windsor, Sir John Thorowgood, of Billingbear, and Mr. Day, of Windsor, added to the Militia Commissions for County Berks."

Vol. IX. Council of State, Day's Proceedings, October 30th. No. 3 : " Commissions to be issued to persons named by Major Evelyn, for command of the Militia forces in Berkshire."

It seems as if the nominations for commissions were entrusted to a committee, of which Major Evelyn was chief. The Regiment was evidently brought up to full strength.

State Papers, Day's Proceedings. December 21st. No. 1 : " One troop of 100 horse to be kept in County Berks."

This troop of horse was probably kept up, or at least the power of calling it out existed, until the 18th Century. That it formed part of the Militia of Berkshire is undeniable, which is curious, as no other parallel case is discoverable.

The bells of Reading were joyously rung on hearing of the prosperous success of the Army in Scotland, when the Scotch were so completely defeated and routed near Edinburgh. The chief towns of Edinburgh and Glasgow at once surrendered to Oliver Cromwell and his victorious army.

A few months later Charles II. was crowned at Scone as King of Great Britain, having signed the solemn League and Covenant demanded by the Scottish people. At the head of his troops he marched into England. But Coronation did not make him a King, for, in the fighting which followed, he was vanquished.

1651.    Colonel Hammond desired to set up a gate and stiles at the entrance into the Forbury, and asked permission of the Corporation to do so. The garrison was still in the Forbury, where it had always been in former days, close to the Fair ground, between the river and the old Abbey ruins.

January 6th. Allowance was asked for the poor maimed soldiers who daily passed through Reading. No provision seems to have been made for the victims of warfare, who were left to private charity when unfit to march with their regiments, ambulances were unheard of, and surgery being of the roughest description.

February 5th. Commissions granted by Council of State : —"Berks : Troop F, Captains Benjamin Burges, Jos. Claver,

John Burningham, Richard Goddard, Nicholas Badcock,
— Doe, John Lush, under Evelyn."

Previous to this, apparently, the troops had been named
from the divisions they were raised in ; but some fresh
arrangement was evidently attempted, as this troop is
distinctly called Troop F. I fancy it was an attempt to
arrange the regiments into brigades.

These eight captains were placed under Major Evelyn ;
perhaps he was captain-lieutenant, and the three officers
named (October 1st, 1650) were the field officers.

March 19th. Vol. XV. Council of State to the Militia
Commissioners for County Berks :

" As the state of affairs now stands, the troop of horse
raised in your county should be continued for the safety of
the county. We, therefore, desire you to keep it up and to
forbear proceeding with further levies upon the Militia, until
order from Parliament or Council.—Whitehall."

August 21st. Vol. XVI. Council of State sent to Lord
Grey and to the Militia Commissioners of Berks the
following :

" You have had notice of the enemy's great marches, and
prepared accordingly. March with all expedition with all
the force you have to the rendezvous, or as directed by
Major-General Lambert or Lieut-General Fleetwood.

" It seems the enemy intends for London, and it therefore
concerns us that they have some check before they come too
near. Give Lieut.-General Fleetwood an account of the
way of your march towards him, and at what places you
intend to quarter and the time of your quartering as near as
you can.—Whitehall."

August 22nd. Council of State to Colonel Whichcott,
Governor of Windsor Castle, and the Militia Commissioners
for Berks :

" We have ordered Captain Cannon to march to
Fleetwood, and, for supply of that force, draw into the

castle one of the foot companies now raised in Berkshire, they being such as you can be assured of, to continue until further orders.—Whitehall."

August 23rd.  Vol. XVI.  Council of State to Lieut.-General Fleetwood :

" By intelligence from the Major-General, we find that the Scots' Army bend their march towards Worcester and Gloucester ; and as it would be inconvenient for the forces of Sussex, Surrey, Hants, and Berks to march to Dunstable, we have given them orders to march to Oxford with all speed, and there to expect your orders  .  .  .  .  .  .  .  .  .  . —Whitehall."

Council of State to the Militia Commissioners of Sussex, Surrey, Hants, and Berks.  To same effect.

August 28th.  Vol. XVI.  Council of State to the Militia Commissioners for Berkshire :

" By yours from Aylesbury, we see your great pains for completing your regiment, and are sorry you meet with those difficulties from the people fit to serve requiring such great sums for bearing arms,—a complaint which has come to us from the other counties, and a mischief which must be provided against hereafter.  For the present, improve all the power the Acts have given you to give the best remedy you can to this mischief.  We like voluntary soldiers if they can be had, rather than men compelled, but leave all to your care to finish that regiment with all expedition.—Whitehall."

On September 3rd, near Worcester, was fought the battle which finally annihilated the hopes of Charles II. and he fled for safety out of his kingdom.

September 13th.  Vol. XVI.  Council of State to the Militia Commissioners for County Berks :

" We have received yours from Wallingford concerning the horses of the Militia troops, to be delivered to the riders who were in actual service upon them, and who charged and did good service there, and have referred the same to consideration ; but you are to adjust it meantime if you can, taking care that none be admitted to such pretensions whose case does not apply, and that no charge grow by continuing them in service until their pretensions are cleared. —Whitehall."

The Berks Militia Horse Troop must have been sent forward and actually been engaged in the Battle of Worcester, that battle was decisive. Charles' Army was scattered and he fled in disguise ; for more than a month he was wandering about the country till at last Shoreham, in Sussex, was reached, whence he crossed the channel in a coal boat to Féchamp, in Normandy.

1653. May. Letters from the Council of State and the Mayor of Marlborough, to the Mayor of Reading and the Marshall-General :

" Sixpence apiece daily was allowed for each one of the Dutch prisoners."

June 7th. A letter was sent to the Commissioners for the Dutch prizes, and for a fortnight's more allowance for the Dutch prisoners. It cost the town of Reading over £65 to keep these prisoners. Another detachment of 100 Dutch prisoners was quartered in Newbury, much to the annoyance of the inhabitants, who resented having to pay for their keep, and were somewhat doubtful whether Parliament would, in the end, refund the expenses.

These prisoners had been captured in the fighting which took place between the English and Dutch Fleets ; a war extending from 1652 to 1654, when peace was concluded.

1655. August 9th. Vol. C. Council of State, Day's Proceedings. No. 10 :

" Major Butler was appointed to command the Militia in the Counties Rutland, Huntingdon, Northampton, and Berks."

1655-56. February 27th. Vol. CXXIV. Council, Day's Proceedings. No. 7 :

" Order, on report from the committee on the business of the Major-General, to advise His Highness to reduce the Militia troops in Counties Oxford, Bucks, Herts, Berks, Hants, Sussex, Kent, Cambridge, Suffolk, Norfolk, and Rutland to 80 in a troop, and pay them off till that time. Approved March 20th."

1658.  On September 3rd, Oliver Cromwell died—the anniversary day of the two greatest victories he had ever achieved.  His rise from the position of a country gentleman to that of Ruler, a king in all but name, was very remarkable ; but his chief power lay as a military commander.  The concluding months of his life were full of troubles.

His son Richard was proclaimed Lord Protector of England, in the Market Place at Reading, September 6th ; but the days of the Republic were over.

Richard Cromwell was not capable of governing the nation : he had neither the will nor the power for such a task.  The Army, under General Monck, took the lead in affairs of State.  Although Richard was proclaimed Protector and acknowledged as such, he allowed the Army to rule him and he was dethroned.

1659.  April 22nd.  From that time England was ruled by her Army and a Council of Officers, Parliament being unable to carry any measures disapproved by them.

July 23rd.  Whitehall.  President Johnson to Major Arthur Evelyn, captain of the Militia troop in County Berks:

" Council learning by yours of the 22nd the condition of the Militia troop under your command, determines to keep it up for 14 days more and no longer, unless you find especial cause ; and if you will send a list of officers and soldiers who have served, they will send a warrant for payment."

August 1.  Warrants for payments by the Council of State.  By J. Frost to Henry Symball :

" For four of Major Evelyn's troopers who brought prisoners from Berkshire to London, £4."

**August 8th.** Vol. CCIV. Index entries of proceedings in the Council of State. Letter to Berkshire Militia. Warrants for payments by the Council of State to Captain Kingdon :

> " These are to will and require you, out of such moneys as are, or shall come, into your hands upon account of the Council's contingencies, to pay unto Henry Symball the sum of four pounds, to be disposed of by him to four troopers of Major Evelyn's troop that brought some prisoners in custody out of Berkshire to London, viz., to each of them twenty shillings. And for so doing this, together with the acquittance of the said Henry Symball, shall be your warrant and discharge.

> " Given at the Council of State at Whitehall this second day of August, 1659.

" JAMES HARINGTON, *President.*

(Signed)  " RO. WALLOP.

" JOHN DIXWELL.

" H. VANE.

" A. JHONSTON.

" To Gualter Frost, Esq., for the Council's contingencies."

**August 10th.** Further instructions were sent for the Militia Horse Troop to be taken care of.

**September 5th.** Council of State. Day's Proceedings :

" Order, since there is now no need for the troop of horse raised in County Berks that the Militia Commissioners there dismiss the troopers to their homes and return the horses to their owners, both to be again forthcoming if wanted."

1659-60.  " January 5th. Berks Militia to pay their men. Berks Militia horse to be returned, and to levy money for the payment of the troops."

**January 29th.** Vol. CCXIX. Council of State to the Militia Commissioners for County Berks :

E

" We think it expedient that the Militia horse raised in your county by the Militia Act, should be returned to their owners until further notice, and you are to levy in a fair and moderate way such moneys as are further leviable by virtue of that Act, for paying the forces there up to this date.— Whitehall."

# CHAPTER V.

## THE RETURN OF THE KING.—1660-1715.

THE return of Charles II. happened in the early part of the year, but his Coronation was delayed till May.

The King was brought back by the Army under General Monck; all the country was weary of the Commonwealth, for, during the ten years it had lasted, it had shewn the disadvantages of a Republican Government. All absence of Court ceremonies; the unsettled and divided feelings of the rival parties; the want of position among Foreign Powers: all these reasons made the people wish again to restore Monarchy, and they welcomed Charles with extraordinary fervour. It was, of course, a critical moment: for if any powerful leader had arisen on the Parliamentary side, after the death of Oliver Cromwell, there must have followed another Civil War.

At once the King set about re-organising the Militia; he also began to form a standing Army, feeling that the Crown needed the support of an Army entirely under its own control. He placed the Militias again under the Lord Lieutenants, a power which had not been enjoyed by them since the reign of Queen Elizabeth, for her successor reserved the supreme control of the Army for himself, although he left the Lord Lieutenants to call out the Militias and arrange all minor details connected with it, under his direction.

Among the State papers of Charles II. reign is found a distribution of the charge of ten Regiments of Horse, consisting of 500 each (besides commissioned officers and non-commissioned officers) with the staff officers, colours,

truinpets, and banners, as it may be raised out of the several
counties, cities and towns yearly, after the rate of 70,000 in
all :

| | £ | s. | d. |
|---|---|---|---|
| The County of Worcester and City of Worcester | 1244 | 8 | 10 |
| The County of Oxford ... · ... ... ... | 1127 | 15 | 6 |
| The County of Berks ... ... ... ... | 1088 | 17 | 10 |
| The County of Gloucester ... ... ... | 1626 | 6 | 8 |
| The City of Gloucester ... ... ... ... | 162 | 11 | 2 |
| The County of Wilts ... .. ... ... | 1944 | 8 | 10 |
| | £7,194 | 8 | 10 |

Also copied from the Calendar, 1660-61, page 560.
March (?)  List of the officers of the said regiments, with
notes of the places where they are to raise their men (by
*Nicholas*) :  (Ninth Regiment ?)  Lord Herbert of Raglan,
Colonel in Gloucestershire ; George Vane, Esq., Major
in Berkshire ; Lord Tracy, Captain in Gloucestershire ;
William Cope, Esq., Captain in Oxfordshire ; Sir Thomas
Doleman, Knt., Captain in Berkshire ; — Erneley, Esq.,
Captain in Wiltshire.

This list is curious, as only two officers of the Berkshire
Regiment are named.  Sir Thomas Doleman was Colonel
of the regiment, and he died in 1711.  The date of his
promotion, however, is not forthcoming.

1665.      War against France was declared in February,
      after several encounters in the Channel between
the two Navies, one of which lasted four days.

July 25th.  The Dutch Fleet was driven from the mouth
of the Thames, and following up the victory, the English
ships carried the war into the enemies' country with fire
and sword.

1666.      July.  In answer to the appeal for soldiers for
      coast defence, Berkshire sent three lusty well-
armed companies, under Major Peacock, which arrived at

the Isle of Wight on July 25th, more being expected to follow.  By July 29th the Berks Regiment at West Cowes numbered 300, under Major Peacock, of whom Lord Lovelace spoke very well ; but the danger was over.   Nevertheless they were not immediately sent back, for, on August 25th, the Trained Bands of the Isle of Wight (Major Peacock's companies from Berkshire and two others) were mustered and were in a good posture.   While the treaty for peace was still being argued out, the Dutch Fleet again attacked the English coasts, this time with success : for they succeeded in passing the fortifications on the river Medway, and, by every tide, were expected to sail up the Thames. Active preparations were made all along the banks of the river, and the local forces were summoned with all speed to arm and proceed to the coasts.   An order in Council was sent to the Lord Lieutenant of Berks, to cause three companies of foot and one of horse to march for the defence of the Isle of Wight, such as had not been drawn the previous year.   The Dutch Fleet, however, altered its course, and made for Portsmouth ; again they failed, nor at Plymouth were they more successful, and once more they sailed for the Thames, getting as far as Tilbury.   The whole English nation was by this time roused to a state of panic.

An Army of 12,000 men was levied hastily.   The crisis was serious, when the announcement of the signing of the Treaty of Breda on July 10th brought peace, and England was saved from invasion.

Lord Clarendon was made the scapegoat to appease his countrymen's rage at the defeat they had suffered and the dangerous position in which England had been placed, through want of expenditure in Army and Navy matters.

1667.   July 27th.   The three companies of trained bands and troop of horse sent from Berkshire to the Isle of Wight, on invasion of the Dutch, had done duty more than a month. The apprehensions of further attempts being over, His Majesty consented, at the instance of Lord Lovelace (the Lord Lieutenant of Berkshire), to their return home ; and authorised him to dismiss them, after examining what pay is due to them for serving beyond the month, and to answer for their debts or borrow money in the Island to pay them, which was to be repaid without delay.

July 29th. From West Cowes, John Lysle to Williamson sent an account of captures at sea and the Dutch advance towards London :

" The drums were beating for the Berkshire soldiers, under the command of Lieut.-Colonel Saunders and Captain Kenrick ; but, upon a letter from Lord Colepepper, that the foreign plenipotentiaries have signed the articles of peace and His Majesty the like, they are ordered to be at Yarmouth, Isle of Wight, to-morrow for their transportation home."

August 5th. No time was lost in sending back the soldiers, for the same correspondent at West Cowes informed Parliament that . . . . . . " The Berkshire and Wiltshire forces departed this day to their own homes."

1673-74. An Address was presented to the King to raise the Militia to show detestation of Popery :

" That the Militia of the City of London and the County of Middlesex may be in readiness at an hour's warning, and the Militia of all the other counties of England at a day's warning, for suppressing all tumultuous insurrections which may be occasioned by Papists or any other malcontent persons."

To this address His Majesty (Charles II.) gave this gracious answer :

" That he would take a special care, as well for the preservation of their persons as of their liberties and properties."

1677. Charles had an Army encamped at Blackheath, under Mareschal Schomberg; many of the officers were Roman Catholics. No wonder the King made preparations to defend himself, remembering as he must have done, the reverses his father had suffered through procrastination and insufficient military organization. Parliament passed the Test Act, but further fighting between the Navies of England and Holland distracted public attention from religious controversies for a while; constant warfare with Holland and France occupied the next few years. The

1678. English Navy was strengthened, and in the spring an Army of more than 20,000 men was raised in a few weeks, all eager to invade France. The Duke of Monmouth was sent to Ostend with 3,000 men, and a Quadruple Alliance was anticipated between the Powers of Europe. All this, however, was frustrated by the Commons, who had not been consulted. The Peace of Nimequen followed, after which home affairs again occupied the King and his subjects.

October 18th. A Petition (among the Reading Corporation M.SS.) was prepared, to be presented to the Duke of Monmouth for easing the Corporation of that town by removing two out of the five companies quartered in the town. The Earl of Clarendon was requested to assist. This petition occurs again on March 5th, 1679. Whether it was ever presented it mattered little, for in November there were still soldiers quartered in Reading. The soldiers on guard were supplied with one bushel of sea coal every night, at 8d. per bushel.

Mr. Powle delivered an address in Parliament, prepared by the Committee appointed for that purpose, most humbly to desire His Majesty that he would command all the Trained Bands to be in readiness, and that one part might do duty for 14 days and after they were dismissed, the two others. And to require them to be very vigilant in the seizing all suspected persons, especially such as travel with arms, or at unseasonable times or in unusual numbers, and likewise to command the Sheriffs to be ready with their posse in cases of insurrection, etc.

It was resolved to send this address to the Lords for their concurrence.

The Lords reply was that they appointed a Committee to consider the Laws relating to the Militia ; who reported that upon inspection of the Statutes they found that, without farther authority, the Militia cannot be kept up above 12 days in one year—thereof four days to be for general musters, and two and two and two and two, viz., eight days for particular musters,—that these 12 days in many counties the Lieutenants have already mustered their men some of these days this present year, not but that by His Majesty's direction (as appears by the Statute) they may be kept up longer. But their Lordships do not find that there is any power to raise the money to pay them.

1684.     The King furthermore promised to recall all commissions given to Papists. At heart it was believed that Charles II. favoured the Roman Catholics ; he certainly had no prejudice against them, and his successor (James II.), who came to the throne in February, 1684, was avowedly of that persuasion, and owed his dethronement as much to his religion as any other cause.

1685.     A Debate on the Militia was held in the House of Commons. By this time these local forces were beginning to require more careful attention. James II., being a Catholic, was viewed with suspicion by his people. They feared a renewal of Papacy, and resisted any attempts at such with energy. The King seemed blind on this point. He carried out his own ideas, without realising the consequences, and he would not yield. General excitement prevailed. The King had every summer encamped his Army at Hounslow Heath to review it, and bring it into proper order and discipline.

A general Muster was called on account of the Duke of Monmouth's invasion, the news of which reached London while Parliament was sitting. They at once promised to support the King with their lives and fortunes, and the Duke was publicly proclaimed a traitor. This declaration was read out in Reading Market Place.

An Army was raised by the King, amounting to 15,000 men. The Militias were called out and warned to be ready for active service.

June 5th. The Duke of Monmouth crossed from Holland with three ships, and landed on the 11th of June on the western shores of England at Lyme, in Dorsetshire, with scarcely a hundred followers; but he was so popular that the country people flocked to join his Standard, and he soon found himself head of an army of over 2,000 men, foot and horse. The Militia of Devonshire assembled for the King, who sent to Holland and re-called six regiments of English soldiers, and drew together all the forces he could to oppose the rebels. Moreover, it was feared the French would aid the rebellion by landing on the South-eastern coast of England. The Battle of Sedgemoor, fought on July 5th, was the last Civil War fought in England. It crushed the rebellion, and on the 15th of July Monmouth was beheaded as a traitor.

James II. had made this rebellion an excuse for retaining his Army in an efficient condition. Every summer the camp at Hounslow was reviewed by him in person.

In 1688 he tried to proselytise the soldiers, a Popish Chapel being erected in the midst of the camp at Hounslow. On June 17th, the very day the Bishops were acquitted, the King had reviewed the troops. He had retired to the tent of the General Lord Feversham, when a great uproar and cheering was heard in camp. The King suddenly inquiring the cause, was told by Feversham, "It is nothing but the rejoicings of the soldiers for the acquittal of the Bishops."

" Do you call that nothing?" replied the King, " But so much the worse for them."

What he meant by this remark does not appear plain, unless severe martial punishment was to follow.

James II. grew more and more unpopular. His subjects at last openly declared for Prince of Orange, and invited him to come to defend the Protestant Religion. Preparations were quickly made by the Dutch; and their old enemies, the French, actually offered to assist the English King against the formidable intrigues impending, which offer was however refused, for James did not believe his son-in-law really had hostile intentions.

The King had relied on his own Army, which numbered some 30,000 regular troops; but these declared against him, with the exception of the Catholic soldiers. In vain he retracted unpopular measures, it was too late. The Prince of Orange's Declaration was openly circulated, and the expedition set sail for the shores of England.

1688.    November 5th. William of Orange landed at Torbay. The Royal Army was still at Hounslow Heath; whether it had remained embodied all the year, or had been re-assembled, does not appear clear. From the Devonshire coast the Prince of Orange marched to Exeter; here he issued his Declaration, inviting supporters to join him. All England was in commotion. The Lord Lieutenants raised the " power of the country "

King James made Salisbury the headquarters of his Army; he hurried thither, and there learnt what a large following had gathered round the Prince of Orange. Day by day fresh desertions were announced. James took fright, retreating towards London. Even his daughter, Anne, withdrew herself from London in her anxiety to escape from politics.

Meanwhile anarchy and confusion reigned supreme. Lord Feversham disbanded the Army without disarming or paying them; this made matters still worse.

A panic arose in Reading, that the Irish Papal soldiers of James II. were pillaging and burning the villages. How this originated no one could say; but somehow the story got about, one told the other, until they believed that the town was on the verge of destruction.

The song, " Lillibullero," was then written, deriding the Papists and the Irish, and it became the party song. 'Tis said by Walpole, in his *New British Traveller*, that the day (December 21st) was kept up yearly afterwards by the inhabitants of Reading, in remembrance of the " Reading Fight."

This skirmish between the Prince of Orange and King James's troops happened at Reading, December 9th (O.S.) The King's Army was in Reading.  The inhabitants, being tired of the expense they had caused, sent out inviting the Prince of Orange and his soldiers to enter the town. The Earl of Feversham was present himself with some Irish Dragoons, and a Scotch Regiment.  They were soon routed and fled from the town in dismay.  Victory was declared for the Prince of Orange.  It is only surprising more were not killed in this engagement, as the townspeople supported the invading force and fired on the King's troops from their windows, who, after their defeat, retired to London, a distance of about 30 miles.  On December 12th the King, in despair, went from London under cover of darkness on board a ship prepared for him.  He only reached Feversham, being there recognised he was compelled to return to London.

William advanced towards the Metropolis.  James remaining at Whitehall, powerless and terrified.  At last the Dutch Guards entered the City and the King was ordered to Rochester.  After a few days he embarked privately for France, on December 23rd ; and William of Orange reigned in his stead.

1690.    Fighting still continued in Scotland and still more seriously in Ireland, where King James set up his Standard, and was not finally defeated till July.

General Musters were held at intervals, one took place in 1697.

The increase of a Regular Army had tended to diminish the importance of the Militia as a National defence.

1697.     A General Peace was signed at Newbourg House, in Holland, where William of Orange had gone to personally arrange the terms of the Treaty of Ryswick. After this, William III. returned to England to attend to home affairs.

December 17. The Commons, among other Bills, appointed several members to prepare and bring in a Bill to regulate the Militias, and make them more useful. They resolved that 10,000 men were necessary for a summer and winter guard at sea for the year 1698.

The French King had a powerful standing Army of skilled soldiers, so that William III. feared to reduce his land forces to any extent ; yet the people had a horror of a standing Army, the bare proposal of such a formation excited all parties. They preferred to re-organise and regularly train the Militias for home defence. They tried to get other regiments disbanded, which had been raised since 1680. At length the sum of £350,000 was allotted for the support of 10,000 men. The King resented this measure, for he did not consider that force sufficient, still he was wise enough not to insist on anything against the wishes of his Ministers. Thus certain troops of Horse Dragoons, and Foot, were caused by the King to be disbanded ; while others were reduced, or sent to Ireland or Scotland, which was done to appease the anxiety of Parliament against a standing army.

1698.     The next year, in December, Parliament again discussed the Army question. The King had not disbanded sufficient troops to satisfy them. The Commons decided that all forces in England, in English pay, exceed-7,000 men, should be at once disbanded ; and that this

Army should consist only of native soldiers, by which clause they aimed at the discharge of the Dutch Guards and the Regiments of French Refugees.

The Army was at once reduced to these limits for guards and garrisons. All William's efforts to retain the services of his faithful Dutch followers were useless, and they were sent back to their native country.

1699. Speeches were delivered by Sir Charles Sedley on various occasions regarding Military affairs. In one speech in Parliament, on the Bill for disbanding the Army, he again declared that 10,000 men he considered enough for defence. Nothing of any special interest was passed that Session with regard to the Militias.

Queen Mary died of smallpox in 1694. William III. survived her eight years ; he died from the results of a fall from his horse at Kensington. Though a prudent ruler, he was not popular as a King, nor was he regretted by his subjects, who, to the last, regarded him as a foreigner.

During Queen Anne's reign England had a peaceful time at home, and there seemed little need of considering the Army. The yearly grants by Parliaments, and Acts regarding the Militias, were passed as usual.

1704. Queen Anne's consent was formally obtained to the Act for better recruiting Her Majesty's land forces and the Marines, for the year 1705 ; also the Act for raising Militia for 1705 (although the month's pay formerly advanced had not been repaid). The Act was also annually passed for punishing mutiny, desertion and false musters, and for better payment of the Army and quarters. This Act was passed every Session down to recent times, because no law existed on the subject; as were likewise the Acts for the paying and clothing of the Militia, which were renewed annually for the same reason.

1707. The hopes of the English Jacobites were raised by the fitting out of a Squadron at Dunkirk, under the Comte Le Forbin, with the intention of placing James Stuart on his father's Throne as James III. The superiority of the English Navy however frightened the French Admiral, who dared not give battle and fled back to Dunkirk.

At that time England was badly garrisoned and Scotland in a still worse condition. Most of the English troops were on the Continent. A panic ensued, and preparations for defence were made. The French had not the pluck to make a more determined effort ; doubtless they felt that it would only mean defeat and a terrible waste of lives for no purpose.

Queen Anne was failing in health. It was thought she favoured the Stuart cause and would nominate the exiled King as her successor. Her death came sooner than was expected, and at the crisis the Hanoverian partisans were prepared ; while the Jacobites had not a sufficiently powerful leader to support the party.

Riots ensued, which in many places in the Midlands and Northern Counties were serious. The Militias were raised ; but the Hanoverians meanwhile crowned George I. as King, and the English people accepted him as their ruler, dreading another war. Besides, the Chevalier St. George had been brought up in France and was supported by the old National enemy, the French King, to say nothing of the Pope. Sir Thomas Dolman, of Shaw, commanded the Berkshire Militia about this time. He died in 1711.

1714.      George I. landed at Greenwich at 6 o'clock, on the evening of September 17th. Among the changes of Ministers, etc., was the dismissal of the Duke of Ormond, and the restoration of the Duke of Marlborough.

The Berkshire Lord Lieutenant, the Duke of St. Albans, was created Captain of the Band of Pensioners. As a descendant of the Stuarts it might have been expected he would have favoured the Jacobites ; but, by accepting a post under the new dynasty, he must have advocated Hanoverian politics.

## CHAPTER VI.

### THE HANOVERIAN DYNASTY.—1715-1757.

THE year 1715 is memorable for the first actual attempt on the part of the Stuarts to regain the Crown of England. The Pretender, otherwise known as the Chevalier de St. George, or, as he desired to be styled, James III., although he had failed in the expedition of 1707, was determined to oppose the Hanoverian succession. His Scotch subjects were eager for rebellion. By the end of August their plans were matured, and the Standard of James III. was raised by the Clans on September 6th, at Aboyne, and within a few days he was proclaimed King in the chief towns of Scotland.

The English Government were aroused. All leaders of the Jacobite faction were promptly seized and imprisoned. Dutch troops were imported to help the English Army; a reward was offered for the Chevalier's head, and every measure was taken to prevent disorder.

Troops were sent to Oxford, the University being avowedly in the Jacobite interest. Bristol was also garrisoned, and such precautions were taken in the West of England, that the Duke of Ormond's attempt to land there was frustrated.

The Militias were embodied in accordance with the preparations agreed upon in the Parliamentary Sessions of July.

October 25th. The Duke of St. Albans, as Lord Lieutenant of Berkshire, issued strict orders to the Constables of his county to bring the Militia up to its full strength and efficiency. Similar directions were sent to every part of the

country. Here again the " Militia Horse Troop" is distinctly
referred to. The Duke's order states :

" That relative to the Acts of Parliament for ordering of
the forces of that part of Great Britain, called England,
I hereby charge and require you, the said High Constables
and everyone of you, forthwith to call together or deliver
precepts to the several petty constables, tything men or
head boroughs of every tything, town, parish, village, hamlet,
or other place within said hundred, directing and requiring
them and everyone of them respectively, by virtue thereof,
forthwith to make return to you in writing fairly written
and subscribed by each of the said officers, making the same
a true and perfect list of the name of the captain who commanded them and the Christian and surnames of the person
or persons who are to find and set forth or contribute to the
finding and setting forth by each foot soldier now to be
found by each and every town, parish, village, tything and
other places within the said Hundred, according to the
number charged upon such town, parish, village, tything and
other place respectively at the last establishment, together
also with the number of soldiers to the Militia horse of the
said county which each such town, parish, village, tything
and other place found and set forth or contributed to the
finding and setting forth at the said last establishment,
and the name of the captain who commanded them and also
the name of all and everyone of the persons who are now to
find, set forth, and contribute to the finding and setting
forth of each and every horse soldier, according to the
number and proportions charged upon each town, parish, &c.,
within the said Hundred at the time of the said last
establishment and the respective places of abode of such
contributories to the horse and where their estates do lie
for which they shall so contribute, and you the said High
Constables are hereby required personally to appear and
deliver all and to everyone the said returns thereof to

me and my deputy lieutenants, at the house of Eleanor Garraway, called the 'Three Tuns' Tavern, in Reading, in the said county, on Saturday, 5th November next, at nine o'clock in the morning, in order for the more speedy and effectual forming into companies and troops and putting in readiness His Majesty's said forces; and you are then and there to attend the service to receive such further orders as shall be thought necessary on that behalf, whereof you are not to fail as you will answer the contrary at your peril.

"Given under my hand and seal this 25th day of October, second year of the reign of George I., 1715."

Robert Lee, Esq., sets out a foot soldier (Edward Dentry, of Binfield). Mr. Jenkinson, Rector, and Mr. Holloway, sets out a foot soldier (John Giles, of Binfield). William Angell, Esq., Mr. Moody, Henry Deane and others, a foot soldier (George Boyce, of Wingfield parish). Richard Hannington, Gent., Mrs. Linscome, Thomas Southey and other contributors, a foot soldier (William Coate, of Oaking-ham). Mr. Toovey, Mr. Thacham, Mr. Lamport and others, a foot soldier (Thomas Puntor, of Binfield). Mr. Baker, Mr. Rhodes, Charles Cowell and others, a foot soldier (James Barns, of Warfield). Mr Barker, Mr. Warneford, Mr. May and others, a foot soldier (Walter Elkins, in the Liberty of Earley, in the parish of Sonning). Sir William Compton, Mr. Ray, Mr. Draycot and others, a foot soldier (Richard Golden, of Binfield). John Pocock, Daniel Pocock, Mr. Griffin and others, a foot soldier (John Turner, of Sandhurst).

The lists for the other divisions, Reading, Abingdon, Newbury, etc., have not been preserved. Each parish had to contribute one or more soldiers, or substitutes, according to its population.

Abingdon at this period was garrisoned by Major-General Pepper's Dragoons, after they had quelled the rioting in Oxford, where they were succeeded by a regiment of foot, under Colonel Handasyde.　　　　　　　　　　F

Meanwhile the Pretender and a small party of his Highlanders crossed the Border, from Penrith they advanced to
Preston, the English Militias of that part of the country
refusing to oppose them.   General Willis was sent against
the invading force, with an Army of six horse and dragoon
regiments, and one battalion of foot, commanded by Colonel
Preston.   The English Army was repulsed at Preston
November 12th ; but when General Carpenter arrived next
day with three regiments of dragoons, the Highlanders were
obliged to surrender.

In Scotland the Jacobite Army was less unfortunate.
The arrival of the Pretender in person at Peterhead, on
December 22nd, again aroused enthusiasm in his cause.   He
was proclaimed King.   A regular Council was formed ; but
want of money and arms, and the severity of the season,
rendered his efforts unavailing.   The English people felt
that the Government was too strong to be opposed : they did
not dare join the rebels.   Sentence of death was passed and
executed upon the imprisoned Jacobite Lords, others
escaped and took refuge abroad.   While on February 4th
the Chevalier deserted his Army, embarked on a small
vessel, and, after a circuitous voyage of five days, landed
at Gravelines.

Thus ended the rebellion of 1715.   Soldiers were still
kept at Oxford, and their presence was much resented.   The
townspeople declared the soldiers to be unruly, and tried to
get rid of them on this plea ; further riots ensued there, but
by degrees the people settled down peaceably again.

In 1717 the House of Commons voted £24,000 for the
payment of four battalions of Munster, and two of Saxe
Gotha soldiers which the King had taken into his service to
supply the place of such as had been, during the rebellion,
drawn from the garrisons of the States General to the
assistance of England ; also they granted £100,000 for the
maintenance of guards, garrisons, and land forces.   The

annual Bill relating to Mutiny and Desertion was also passed. The standing forces were, in the winter Session of this year, fixed at 16,347 effective men.

1718. The number decided upon for land forces was 12,435 men. Although no direct mention is made of Militia musters, I feel convinced some training took place annually, for the land forces above decided upon could not have been composed entirely of regular regiments.

1719. The South Sea Scheme occupied all men's minds to the exclusion of other topics for several Sessions, till the crisis came in 1720, and ruin fell upon the unhappy speculators.

The Mutiny Bill exercised Parliament greatly during the reign of George II. (1720-9), and the discipline of the Army was evidently little or *nil*, as is shewn by the horror evinced at the idea of a standing army in times of peace.

1726. The Militia of England was commuted at 200,000 horse and foot soldiers. They mustered once or twice yearly. The horsemen being allowed during time of muster 2s. per day, and the foot soldiers 1s. daily. The Lord Lieutenant had full power over his regiment, both as to levying taxes to pay it (according to the rate fixed by Act of Parliament), and also as to calling out and arming the soldiers.

1733-34. Debates took place in Parliament about the removal of officers. By the two Militia Acts, passed 13th and 14th (Charles II.), the King could remove officers of Militia. The Militia were pronounced of no use, after other nations began to keep standing Armies, and there was some truth in the assertion. Yet the House of Commons still steadily resisted a regular Army being established, and so long as there was no regular Army, England had only the Militia to rely upon.

In the House of Commons, Mr. Andrews (the Deputy Paymaster of the Forces) moved for an addition of 1,800 to

the land forces, such as had been added the previous year,
when the unsettled state of European politics drew attention
to the condition of the English Army.

1734.    The land forces were again augmented by the
King and were estimated at 27,744 effective men.

Europe was in a state of wars and rumours of wars. Yet
in the following year the English forces were reduced, for
economical reasons and national prejudice.

1735.    The Porteous Riots in Edinburgh, and other
questions relating to smuggling, occupied Parliament; but in 1736 the land forces numbered 17,704
effective men.

1740.    In a debate on the Army in the House of Lords,
the Duke of Argyle expressed his opinion that the
Army ought to be under the control of one man. Com-
plaints were made of the want of discipline. The way the
levies of previous years had been tyrannically administered,
and Colonels appointed unsuitable and bad; these were
serious matters, and it had to be considered whether it was
advisable to raise new regiments to increase the force or
add new men to the old existing regiments. The chief
difficulty being a question of pay.

Our Army as compared with other nations, especially the
French Army, was at a disadvantage, the latter for years
had had a standing Army of veteran soldiers; and the
raising of new regiments was condemned as dangerous to
the peace and welfare of England, yet it was felt on all sides
that some effort was necessary for the safety of the United
Kingdom.

1740 was one of the years celebrated for a long frost.
It began at Christmas and continued till the end of
February. The Thames was thickly frozen over. The cold
was terrible, and many persons died. Prices rose, and
poverty among the poorer classes was cruelly felt. Preparations for war were being carried out by the English

Ministry. Camps were formed at Hounslow Heath, the Isle of Wight, Colchester, and in other parts.

Secret meetings were taking place among the Scotch Jacobites, who were only waiting for arms and money to again rise in rebellion.

The Duke of Newcastle spoke on this same question of the Army; as did also Lord Carteret of the troops of Ireland, Sweden, Hanover, and France as compared with England. The Earl of Cholmondeley and other noblemen also spoke on the Army, each apparently against the establishment of a standing Army. The motion for rejection: Content, 42; not content, 59. A protest was entered, to this effect: one point being, that as new regiments had been proved more expensive and as a new Parliament was so near, no additional regiments had better be raised. And it was alleged the fate of the war had been left to raw new levied troops, in order to keep others at home only for evil purposes.

1741. The King determined to send troops to the Netherlands, and, by this unwise move, involved England in the Continental troubles. Of this war, the chief

1743. remarkable point was that the Battle of Dettingen was the last time an English King led his own army into the field of battle.

Soon afterwards the British Army returned from the the Netherlands.

The English now made a protest against the Hanoverian troops in British pay, Parliament was determined to economise by disbanding them.

An expected invasion of the French, in support of the Stuart family, again threatened Britain. Regiments were sent to the South coast. Governors and Commanders were · ordered to their posts. The forts on the Thames and Medway were strengthened, and the Kent Militia was called out.

January.  Troops were raised and collected ; the French preparations to support the Young Pretender causing general alarm.  The Jacobites were only waiting for an opportunity to rise in favour of the Stuart family ; meanwhile they were holding secret meetings and planning an invasion.  While Howard's Regiment of Foot was being recruited for service in Flanders, a fine young woman of 5ft. 6in., in man's clothing, came to the Sergeant in Reading to be enlisted.  Before they left the town, however, her sex was discovered ; but she refused to return the marching guineas, as she had spent half.

1744.       On March 31st, England declared war against France and France against England.  Thus hostilities commenced.  An Act of High Treason was passed against the Stuart family and all who supported their cause.

1745.       July 25th.  The long threatened invasion of the Stuarts at last came to a crisis.  Prince Charles in person landed in the Isles on the northern shores of Scotland, at a moment when most of the English troops were away serving in the Netherlands.

By degrees the chieftains gathered round his Standard.  He was proclaimed Prince Regent, under his father, King James VIII. of Scotland and III. of England.  After an extraordinary victory over the English Army at Preston Pans, the Scotch Clans were led southwards across the Border to the conquest of England.  So sudden and unexpected was this advance that everyone fled before them, and without opposition the Rebel Army marched through Northumberland, Cumberland and Lancashire, moving towards Cheshire, where Jacobitism had been actively kept up.  Few recruits, however, joined the Army ; and the noblemen of those parts could not realise that the Prince had actually arrived in England.  The march continued until Derby was reached on December 4th.  King George's Army, under the Duke of Cumberland, then being only nine miles distant at Newcastle-under-Lyne.

A large Army from France was prepared to effect a landing on the South coast. The whole of the English nation was aroused and frightened into activity. Militias and Trained Bands were called out. A camp of 5,000 or 6,000 troops occupied Greenham Common. Curiously, the old *Reading Mercury* of that date, although giving an account of the movements of the rebels, says little or nothing of the efforts made locally; perhaps the editor was a Jacobite! The King assembled the bulk of his Army at Finchley; while, as on previous occasions, regiments were sent to guard the southern and western coasts. A sudden panic seems to have seized the Jacobite leaders. They were disappointed at the lack of support given them in England, and, with one accord, insisted on a speedy return to Scotland. In vain, Charles Edward urged them to advance on London: he was powerless to influence them, and the fatal retreat began on December 6th. If only they had delayed a few days, support would have come. Messengers had been sent to Derby from several powerful noblemen, among others Sir Watkin Wynn and Lord Barrymore of Hurley; but too late to be of any use.

On the 8th of November, 1745, the Jacobite Army had first entered England. The only fighting took place on their retreat at Clifton Moor, not far from Penrith, and it resulted in victory for the Highlanders, who, however, still continued their retreat northwards, re-crossing the Border on December 20th, closely pursued by the Duke of Cumberland. The Battle of Culloden Moor took place on April 15th. Prince Charles Edward was obliged to disguise himself, and, after many adventures and hairbreadth escapes, sailed for France on September 20th. It is said that several years afterwards he re-visited England *incognito*, but all hope of regaining the Crown was at an end. There is a tradition, for which no foundation can be traced, that he visited Berkshire secretly.

*Gentleman's Magazine*, May, 1746. Vol. XVI., page 237:
"In a Bill for the regulation of the Militia, the number of
private foot soldiers to be annually raised in the counties of
England as follows: Berks, 700; &c."

1746. After the failure of this rebellion, the land forces
were 60,000, including 11,500 Marines. Some
regiments of Life Guards were disbanded for economical
reasons, and other regiments reduced.

In March, 1746, John Heddige was tried at the Assizes
in Oxford for desertion in the late rebellion.

1748. War still continued on the Continent until the
articles of a general peace were signed at Aix La
Chapelle on October 7th.

The land forces of England again occupied the consideration of Parliament; one question being whether half-pay
officers were subject to martial law. The new Mutiny Bill
and the old fear of a permanent Army was again discussed.
The land forces, as in the previous year, were continued at
18,857 men.

Peace was publicly proclaimed in London, on February
2nd, 1749.

1750. This year was chiefly remarkable for storms,
thunder, wind, hail, and rain. A strong shock of
earthquake was felt in February in and about London, and it
was prophesied that the end of the world was near at
hand. The death of the Prince of Wales was another
calamity of this year, which came as a sudden disaster on
the nation.

1752. The year 1751 passed without event; but 1752
will ever be remembered as the year in which
eleven days were taken out of the calendar, and from
September 2nd the following day was reckoned as 14th of
the month, so that the year was thus made to commence on
January 1st, not March 25th, as it had formerly done in
England.

1753. *Universal Magazine Supplement,* Vol. XXII, gives "An abstract of an Act to explain, amend, and enforce an Act made the last Session of Parliament, entitled an Act for the better ordering of the Militia forces in the several counties of England, an extract from which is inserted in our Magazine, Vol. XX., p. 322." The extract is too long for me to copy, but it gives the following points : —First, that Militias shall be raised at once, where not already done. Certain estate or income qualifications, a twenty-one years' lease of an estate or £300 a year, being the minimum for officers. Every captain might appoint or displace corporals and drummers, and sergeants, to fill vacancies ; but the colonel or commanding officer could remove such sergeants. Lists of inhabitants, to be made between the ages of eighteen and fifty, to be placed on church doors. Enrollment to be for three years, with a penalty of £10 for not serving. No peer, &c., was liable to serve personally. Men ill on march were to be provided for, &c. Allowances were to be made for families, &c. Levies, &c. Training and pay were all discussed with a view to improvement.

1755. Two years later another article in the same Magazine is to be found, on the flourishing state of Great Britain. It refers to the question of a standing Army being necessary, as such are kept on the Continent, especially France. It compares the Militia with a standing Army and the organisation of the former, so as to avoid jealousy. Evidently, from the wording, there was jealousy between the Militia and the Army, as it alludes to the opinions of "those who urge that the officers of the Army ought not to be admitted into the Militia." The Militia bounty money was higher than in the Army, and later on the payment of substitutes was raised, as also the allowances for the wives and families of men serving. Those points all rendered the Militia the more popular service of the two ; for, whereas the Government paid for the Army, the Militia was to a great extent supported by the liberality of its officers, who prided themselves on the efficiency of their regiments and did all in their power to keep them up to a high standard.

1756.      In the Winter Session of Parliament, a new Bill was prepared and brought in, for the better regulating of the Militia forces of the English counties. It was passed by the Commons on May 10th, but opposed by the House of Lords. Nothing further was done until another threatened French invasion was being prepared. Whilst Parliamentary debates were being carried on, as to whether the Militia forces were sufficient to protect England, George II. sent for a body of Hessian and Hanover troops, which duly arrived and were encamped in different parts of England.

On February 3rd, a Proclamation was issued. All officers, civil and military, were commanded to watch the coasts, and, on the first sight of the enemy, to announce the same by beacons and the beating of drums. All articles of food and all cattle were to be taken away twenty miles inland. Bridges were to be broken down and roads barricaded, to hinder the advance of a hostile force if it succeeded in landing. These orders are almost identical with those issued 200 years before by Henry VIII.

# CHAPTER VII.

PROBABLY the arrival of the German troops contributed more than anything to the improvements which shortly after took place in the English Army. New regiments were raised and the old ones re-modelled ; recruiting was actively carried out. The number of private men to be raised by each county was fixed. A total of Militia in England gave 32,000 men. The Lord-Lieutenant of Berks was the Duke of St. Albans, and the number of men to be raised in the Royal county was fixed at 560.

" Where a town lies in two counties, the inhabitants shall serve in the Militia of that county in which the church is situated. Workingham (? Wokingham) is to join the Militia of Berks."

In the Army List of 1762 it is stated the Berkshire Militia were embodied July 25th, 1757. Unless this date is wrongly given, the regiment was embodied before that of other counties, whether it was 1757 or 1758. At any rate, the Berkshire Militia were among the first called to arms ; a fact worthy of record, considering the difficulty experienced in some parts of England.

1758. December 10th. Militia Letter Book (204) Record Office, Whitehall :

" *To the Duke of St. Albans.*

"My Lord,—Having laid before the King the list transmitted to me by William Brookland, Esq., Clerk of the General Meeting for the County of Berks, of the names of the several gentlemen who have offered themselves to serve as officers in the Militia for the said county. I have His Majesty's commands to acquaint your Grace that the King does not disapprove of any of the gentlemen therein mentioned, to be officers in the Militia of the said County of Berks.—I am, &c., HOLDERNESSE."

" The names of the several gentlemen who have offered themselves to serve as officers in the Militia for the County of Berks:—Sir Willoughby Aston, Baronet, Colonel; Arthur Vansittart, jun., Lieut.-Colonel; John Dodd, Major; John Blagrave, Captain; Thomas Draper Baber, Captain; William Mackworth Praed, Captain; Joseph Andrews, Captain; Richard Sellwood, Captain: John Reeves, Lieutenant; John Walter, Lieutenant; John Wilder, Lieutenant; Lawrence Head Osgood, Lieutenant; Peter Floyer, Lieutenant; Edmond Seymour, Lieutenant; Thomas Justice, Lieutenant; George Hatch, Ensign; Thomas Buckeridge Noyce, Ensign; John Boult, Ensign; William Towsey, Ensign.
" Examined by me the seventh day of December, 1758.

"WILLIAM BROOKLAND,
" Clerk of the General Meeting for the County of Berks."

The following extract is from a pamphlet entitled, " Reflections on the different ideas of the French and English in regard to cruelty." The humble petition of the prisoners in the —— for debt, addressed to the humanity of the nation in Parliament assembled :

."  .   .   .   .   . at home we are shrinking into depopulation to a degree melancholy to those who observe and reflect. What difficulties do we not find in raising the few forces necessary to our defence! Can anything but the absolute scarcity of men account for the incompletion of our Militia? For it cannot be supposed that we, everyone of us, wished for it with so much zeal and ardour, with an exception of our own particular service in it. A kingdom like this, when it has not a hundred thousand men to spare upon an emergency, is an estate that can only make shift to support its owner until he has a fit of sickness, but then cannot pay the doctor's bill.  .   .   .   ."

*From a portrait at Shaw House.*

Sir JOSEPH ANDREWS, Bart.
1757.

1759. March 5th. The return of the officers, &c., for the Militia Regiment for the County of Berks.

*" To the King's Most Excellent Majesty.*

" We, whose names are hereunto set and subscribed, three of the Deputy Lieutenants of the County of Berks, in the absence of George Duke of St. Albans, your Majesty's Lieutenant for and in the said County of Berks, do hereby humbly certify and return to your Majesty that, in pursuance of two several Acts of Parliament made and passed in the thirtieth and thirty-first years of your Majesty's reign, the one intituled ' An Act for the better ordering of the Militia forces in the several counties in that part of Great Britain called England,' and the other being ' An Act to explain, amend, and enforce the said first mentioned Act.' Three-fifths of the Militia men of the said County of Berks have been chosen, sworn, and enrolled ; and that three-fifths of the commission officers of the Militia forces raised, and to be raised in and for the said county have been appointed, taken and out, and accepted their several qualifications, as by the said several Acts of Parliament in that case made, they are directed, viz. :— Field Officers : Sir Willoughby Aston, Bart., of Wadley ; Arthur Vansittart, Esq., of Shottesbrooke ; John Dodd, Esq., of Swallow-field. Captains : John Blagrave, Esq., of Southcote ; Thomas Draper Baber, Esq., of Sunninghill Park ; William Mackworth Praed, Esq., of Warfield ; William Andrews, Esq., of Shaw ; Richard Sellwood, Esq., of Peasmore. Captain Lieutenant : Thomas Blagrave, Esq., of Lambourn. Lieutenants : John Reeves, Esq., of Arborfield ; John Walter, Esq., of Swallowfield ; John Wilder, Esq., of Nunhide ; Lawrence Head Osgood, Esq., of Barkham ; Peter Floyer, Esq , of Shinfield ; Thomas Justice, Esq., of Sutton Courtney ; Edmund Seymour, Esq., of Lambourne Woodlands. Ensigns : George Hatch, gent., of New Windsor ; John Dean, gent., of Ruscomb ; Thomas Buckeridge Noyes, gent., of Southcote ; John Boult, gent., of Charridge ; James Pettit Andrews, Esq., of Shaw ; William Towsey, the younger, gent., of Wantage.

" Dated March 5th, 1759.

" WILLOUGHBY ASTON.
" JOHN DODD.
" ARTHUR VANSITTART."

Militia Correspondence. Record Office, Board of Ordnance, Whitehall, April 4th, 1759:

"Gentlemen,—Three of the Deputy Lieutenants of the County of Berks, in the absence of His Grace the Duke of St. Albans, His Majesty's Lieutenant for and in the said County of Berks, having, agreeable to the Acts of Parliament for the better ordering of the Militia forces in that part of Great Britain called England, certified and returned to the King that three-fifths of the Militia forces for the said County of Berks have been chosen and enrolled ; and that three-fifths of the commission officers for the same have been appointed and taken out their commissions, and entered their qualifications ; and they, the three Deputy Lieutenants, having in consequence thereof desired that the necessary arms, accoutrements, &c., may be delivered for the use of the said Militia, I am to signify to you His Majesty's pleasure that you do accordingly direct the arms and accoutrements agreeable to the list enclosed, to be provided and delivered free of any expense of carriage, at such places in the County of Berks as the Duke of St. Albans, His Majesty's Lieutenant thereof, or the three Deputy Lieutenants aforesaid shall judge most convenient, and to such person or persons as shall be duly authorised by them or either of them to receive the same.—I am, &c., HOLDERNESSE."

"A return of arms, accoutrements, and ammunition necessary for the Militia for the County of Berks, commanded by His Grace the Duke of St. Albans, consisting of 30 sergeants, 20 drummers, and 560 rank and file, formed into eight companies:—Silk colours, the one an Union, the other a blue sheet with the arms of His Grace the Duke of St Albans, 2 ; oilskin cases, lined with bays (baize), 2.  Rank and file: Muskets with wood hammers, bayonets, scabbards with tanned leather slings, 560 ; cartouch boxes, with belts and frogs, 560 ; brushes and wires, 560 ; small hangers with brass hilts, scabbards, and tanned leather waist belts, 560 ; iron wiping rods with worms, 30. Sergeants : Halberts, 28 ; large hangers, with brass hilts, scabbards, and tanned leather waist belts, 28. Drummers : Foot drums complete, with the arms of His Grace the Duke of St. Albans, 20 ; drum sling or carriages, 20 ;

ticken drum cases, 20 ; hangers with brass hilts and scabbards, including one for the drum-major, 21 ; waist belts the same as the drum carriages, 21. Ammunition : Powder—service, 4 barrels ; exercise, 4 ; musket-ball, 8 cwt. ; flints, 2,240 ; formers, 30 line paper, 4 reams ; leather powder bags, 30.—H. BEAUCLERK."

In June, 1759. the Company of Berks Militia, commanded by John Dodd, Esq., Major, was drawn up at Whitley Wood, near Reading, where they were exercised for the first time, and received their arms, clothing, &c.

The Militia of several counties were reviewed that month by their commanding officers, in the presence of the Lord Lieutenants and a great number of persons of distinction. They all performed their exercise amazingly well, behaved dutifully to their superiors, soberly in their quarters, and seemed full of cheerfulness and alacrity, and ready to march wherever they were ordered for the defence of their country. An order in Council was published on July 11th, declaring " that all His Majesty's faithful subjects who enlisted themselves in the land service from that day, should not be sent out of Great Britain, and should be entitled to their discharge at the end of three years or at the end of the war, whichever they preferred, and all deserters who rejoined before August 20th their respective regiments or any other corps, if their own were out of the kingdom, should be pardoned."

STATE OF THE MILITIA OF THIS KINGDOM.

*Universal Magazine*, July : " Berks, Lord Lieutenant, Duke of St. Albans. Number to be raised, 560. Officered and near completion, 560 now on duty."

Several counties had then their full strength and are reported as on duty, while in other parts neither men nor officers could be found willing to serve.

Marching Orders :

" It is His Majesty's pleasure that you cause the Berkshire Regiment of Militia under your command to assemble with all

convenient speed at such place or places as you shall think proper,
and march from thence by such routes and in such divisions as
you shall think most convenient to Marlborough, Hungerford, and
the Devizes, where they are to be quartered, and follow such orders
as they shall receive from Major-General Holmes.—Given at the
War Office this 26th day of July, 1759. By His Majesty's command, BARRINGTON."

This probably was in consequence of a report that the
French had already landed at Dover, which arose from some
strange ships being sighted off the coast. Special messengers
were sent to London and the whole country prepared for
active defence. The land forces of Great Britain consisted
of two troops of Horse Grenadier Guards, seven regiments
of Dragoons, three regiments of Foot Guards, thirty-four
regiments of foot, thirty-two independent companies, and in
Ireland four regiments of horse (six of dragoons and twelve
foot regiments).

" To His Grace the Duke of St. Albans, His Majesty's Lieutenant
    for the County of Berks, or, in his absence, to any three
    Deputy Lieutenants for the said county.
    " It is His Majesty's pleasure that you cause two companies of
the Berkshire Regiment of Militia, under your command at
Hungerford, to march from thence by the shortest and most
convenient route to Marlborough and Preshute, where they are to
be quartered and remain until further orders —Given at the War
Office this 29th day of August, 1759."

" To Lieut.-General Holmes, or officer commanding the two corps
    of the Berkshire Regiment of Militia at Hungerford.
    " It is His Majesty's pleasure that you send such detachments
of the Berkshire Regiment of Militia at Marlborough, as you shall
think proper from time to time, to the Devizes, where they are to
be quartered and remain until further orders.—Given at the War
Office this 29th day of August, 1759."

"To Lieut.-General Holmes, or officer commanding the Berkshire Regiment of Militia at Marlborough."

"It is His Majesty's pleasure that you cause the Regiment of Berkshire Militia under your command to march immediately from their present quarters by the shortest and most convenient route to Winchester, where they are to be quartered in the barracks and remain until further orders.—Given at the War Office this 25th day of October, 1759."

*Universal Magazine*, 1760, gives a song called

### "ON THE MILITIA'S EXPEDITION."

" Bellona spreads her dire alarms
And calls the Britains forth to arms ;
With eager haste behold them fly,
Resolved to conquer or to die.

With joy the glorious call obey,
For glory points them to the way ;
Undaunted they their foes will meet
And triumph over Gaul's defeat.

Let dastard foes be aw'd by fear
And tremble when no danger's near ;
The gallant heart no danger knows,
But pants to meet great George's foes.

Britannia raised her drooping head
And smiling thus, the Goddess said—
' My sons, the glorious task pursue,
Maintain your rights and France subdue.'"

The Commons passed a Bill to enable His Majesty's Lieutenants of the several counties, ridings, and places, to proceed with the execution of the laws relating to the Militia, notwithstanding any suspension of the same, and for other purposes relating to the said laws. All provisions of the Mutiny Act extended equally to Militia.

1760.  June.  Regulations were issued as to officers in each regiment, and the maintenance of Militiamen's families, while the regiment was embodied.

G

" To Lieut.-General Holmes, at Portsmouth.

" It is His Majesty's pleasure that (notwithstanding any former order to the contrary) you cause the Berkshire Militia under your command at Winchester to march from their present quarters on Tuesday, the 17th instant, to the place of encampment near Winchester, where they are to encamp and remain until further orders.—Given at the War Office this 13th day of June, 1760."

There were, besides the Berkshire Militia at this encampment, the Militias of Gloucestershire, Bedfordshire, Dorsetshire, and Wiltshire, and the 34th (Buckinghamshire) Regiment. The Earl of Effingham was in command, the Earl of Shaftesbury Brigadier-General, and Edward Montagu (of the Wiltshire Regiment) Major of Brigade.

1st Division: Hungerford and Ilsley.

2nd Division: Newbury and Speen.

" It is His Majesty's pleasure that you cause the Berkshire Regiment of Militia, encamped under your command near Winchester, to march from thence in two divisions, according to the routes annexed to the places mentioned in the margin hereof, where they are to be quartered and remain until further orders.—Given at the War Office this 4th day of October, 1760."

" To Earl of Effingham, or other officer commanding the forces encamped near Winchester. Route for the 1st Division of the Berkshire Regiment of Militia from Winchester camp : Thursday, October 9th, Andover ; Friday, October 10th, Newbury ; Saturday, October 11th, Hungerford and Ilsley, where they are to remain. Route for the 2nd Division of the Berkshire Regiment of Militia from Winchester camp : Friday, October 10th, Andover ; Saturday October 11th, Newbury and Speen, where they are to remain.

" It is His Majesty's pleasure that (notwithstanding the order of the 4th instant) you cause the Berkshire Regiment of Militia, encamped under your command near Winchester, to march from thence in two divisions, according to the routes annexed to the places mentioned in the margin thereof, where they are to be quartered and remain until further order.—Given at the War Office this 8th day of October, 1760."

"To the Earl of Effingham, or officer commanding the forces encamped near Winchester. Route for the 1st Division of the Berkshire Regiment of Militia from Winchester camp : Thursday, October 9th, Whitchurch and Overton ; Friday, October 10th, Newbury ; Saturday, October 11th, Hungerford and Ilsley, where they are to remain. Route for the 2nd Division of the Berkshire Regiment of Militia from Winchester camp : Friday, October 10th, Whitchurch and Overton : Saturday, October 11th, Newbury and Speen, where they are to remain."

"It is His Majesty's pleasure that you cause the Regiment of Berkshire Militia under your command to march immediately from their present quarters, by the shortest and most convenient route, to the places mentioned in the margin hereof (acquainting this office of the day on which they begin their march and the day on which they will arrive at their destined quarters), where they are to be quartered and remain until further order. —Given to the War Office, this 20th day of November, 1760. By His Majesty's command, in the absence of the Secretary at War, THOS. TYRWHITT."

Companies :
Reading ... 5
Wallingford 2
Oakingham 2
—
9

Sent by express at 3 o'clock, on Friday, Nov. 28, 1760.

"To Colonel Sir Willoughby Aston, Bart., or officer commanding the Berkshire Regiment of Militia at Newbury. It is His Majesty's pleasure that you cause two companies of the Berkshire Regiment of Militia under your command at Reading to march from thence immediately, with all possible expedition, by the shortest and most convenient route to Witney, in Oxfordshire, where they are to be quartered and remain, and be aiding and assisting to the civil magistrates, to follow such directions as they shall receive from time to time from the said civil magistrates, for suppressing any riots or tumults which may arise in that neighbourhood, and in securing the rioters and preserving the public peace ; but not to repel force with force, unless it shall be found absolutely necessary, or being thereunto required by the civil magistrates.—Given at the War Office this 18th day of March, 1761."

"To Colonel Sir Willoughby Aston, or officer commanding the Berkshire Militia at Reading, Berks. It is His Majesty's pleasure that you cause the companies of the Berkshire Regiment of Militia under your command at Reading to march from thence on Tuesday, the 14th instant, in such divisions as you shall think proper (provided the several places through which they are to march, and in which they are to be quartered on their march, be not within a less distance than two miles from any town or city, where the election for Members of Parliament shall be held within the time of their march, and acquainting this office of the day they will arrive at their destined quarters), to any other place or places in the County of Berks, except Newbury and the Parish of Speen, where they are to be quartered and remain until further order. — Given to the War Office this 3rd day of April, 1761. CHAS. TOWNSHEND."

"To Officer commanding the companies of the Regiment of Berkshire Militia at Reading. It is His Majesty's pleasure that you cause the two companies of the Berkshire Regiment of Militia under your command at Witney to march from thence, by such route as you shall think most convenient (provided the several places through which they march, and in which they are to be quartered during their march, be not within a less distance than two miles of any town or city where the election for Members of Parliament shall be held during the time of their march, and acquainting this office with their disposition), to such place or places in the County of Berks as you shall think proper, except Reading, Newbury, and the Parish of Speen, where they are to be quartered and remain until further orders.—Given at the War Office this 7th day of April, 1761."

"To Officer commanding the companies of the Berkshire Militia at Witney, Oxfordshire. It is His Majesty's pleasure that (notwithstanding any former order to the contrary) you cause the Berkshire Regiment of Militia under your command to march from their present quarters by the shortest and most convenient route, in such divisions as you shall think proper, to such place or places in the County of Berks (except Reading, Newbury, and the Parish of Preshute as you shall think proper), providing the several places through which they march, and in which they are to be quartered during their march, be not within a less distance than two miles of any town or city where the election for Members shall be held during the time of their march, and acquainting this office with their destined quarters, where they are to be quartered and remain until further orders.—Given at the War Office this 10th day of April, 1761."

"To Colonel Sir Willoughby Aston, or Officer commanding the Berkshire Regiment of Militia. It is His Majesty's pleasure that you cause the Berkshire Regiment of Militia under your command to march from their present quarters, by the shortest and most convenient route (acquainting this office of the day they will arrive at their destined quarters), to Reading, where they are to be quartered and remain until further orders."

"To Colonel Sir Willoughby Aston, Bart., or Officer commanding the Berkshire Regiment of Militia at Abingdon. It is His Majesty's pleasure that you cause His Majesty's Regiment of Militia for the County of Berks under your command, to march from their present quarters in two divisions, according to the routes annexed, to the place of encampment near Winchester, where they are to encamp and remain until further order."

"To Colonel Sir Willoughby Aston, Bart., or Officer commanding His Majesty's Regiment of Militia for the County of Berks at Reading. Route for the 1st Division of His Majesty's Regiment of Militia for the County of Berks, from Reading to Winchester camp: Friday, June 12th, Basingstoke; Saturday, June 13th, Suttons, Nuttley, and places adjacent; Sunday, June 14th, halt; Monday, June 15th, Winchester camp, and remain. Route for the 2nd Division of His Majesty's Regiment of Militia for the County of Berks, from Reading to Winchester camp: Saturday, June 13th, Basingstoke; Sunday, June 14th, halt; Monday, June 15th, Suttons, Nuttley, and places adjacent; Tuesday, June 16th, Winchester camp and remain."

1761. In 1761, there were encamped near the site of the Hessian camp the Wilts and the South

Battalion of the Gloucestershire Militia, the Dorsetshire and the North Battalion of the Gloucestershire Militia, and the South Battalion of the Hampshire and the Berkshire Militia.

"To Colonel Sir Willoughby Aston, Bart., or Officer commanding His Majesty's Regiment of Militia for the County of Berks at Sutton. It is His Majesty's pleasure that you cause the quarters of the two divisions of His Majesty's Regiment of Militia for the County of Berks under your command, on their arrival at Sutton, Nuttley and adjacent places to be enlarged with [? near] Overton. —Given at the War Office this 6th day of June, 1761. By His Majesty's command, in the absence of the Secretary at War, and his deputy, THOS. BRADSHAW."

" To Lieut.-General Earl of Effingham, or Officer command- ing the forces encamped near Winchester. It is His Majesty's pleasure that you cause His Majesty's Regiment of Militia for the County of Berks under your command, to march from the place of encampment near Winchester, on Tuesday, the 20th instant, in two divisions, the 2nd division the day after the 1st, by the shortest and most convenient route (acquainting this office with their route and the day they will arrive at their destined quarters), to the places mentioned in the margin hereof, where they are to be quartered and remain until further orders.—Given at the War Office this 15th day of October, 1761."

Reading, Wallingford, and Oakingham.

"To Lieut.-General the Earl of Effingham, or Officer com- manding the forces at Winchester. It is His Majesty's pleasure that, notwithstanding any former order to the contrary, you cause His Majesty's Regiment of Militia for the County of Berks under your command to march to-morrow, being Tues- day, the 20th instant, from the place of encampment near Winchester, by such routes and in such divisions as you or the officer commanding the said regiment shall judge most con- venient (acquainting this office with their route and the day on which they will arrive at their destined quarters), to Reading, where they are to be quartered and remain until further orders. —Given at the War Office this 19th day of October, 1761. By

His Majesty's command, in the absence of the Secretary at War, THOS. TYRWHITT."

" To Colonel Sir Willoughby Aston, Bart., or Officer commanding His Majesty's Regiment of Militia for the County of Berks at Reading. It is His Majesty's pleasure that you cause your companies of His Majesty's Regiment of Militia for the County of Berks under your command at Reading to march immediately from thence, by the shortest and most convenient route (acquainting this office with the day of their arrival at their destined quarters) to Newbury, where they are to be quartered and remain until further orders.—Given at the War Office this 11th day of November, 1761. By His Majesty's command,

"CHAS. TOWNSHEND."

ARMY LIST, MILITIA, 1762.

THE BERKSHIRE REGIMENT OF MILITIA, EMBODIED JULY 25TH, 1757.

| Rank. | Name. |
|---|---|
| Colonel | Sir Willoughby Aston, Bart. |
| Lieutenant-Colonel | Arthur Vansittart. |
| Major | John Dodd. |
| Captains | William Mackworth Praed. |
| | Joseph Andrews. |
| | Richard Sellwood. |
| | Clement Saxton. |
| | Laurence Head Osgood. |
| | John Wilder. |
| Captain-Lieutenant | Thomas Blagrave. |
| Lieutenants | John Reeves. |
| | John Walter. |
| | Peter Floyer. |
| | Edmund Seymour. |
| | Pettit Andrews. |
| | Thomas Justice. |
| | Samuel Southby. |
| | Thomas Baber. |
| | Thomas Buckridge Noyes. |

|           |                        |
|-----------|------------------------|
| Ensigns - - - - - - | Thomas Grove.<br>John Fortescue Alland.<br>Henry Evans.<br>Clement Styles.<br>William Towsey.<br>Joseph Langford.<br>Francis Annesley. |
| Adjutant - - - - - - | Henry Evans. |
| Quartermaster - - - - | Thomas Buckeridge. |
| Surgeon - - - - - - | John Fortescue Alland. |

Agent - - - - - - Mr. Pye, Featherstone Buildings.

" To Colonel Sir W. Aston, or Officer commanding the Berkshire Militia at Reading, Berks.

" It is His Majesty's pleasure that you cause His Majesty's Regiment of Militia for the County of Berks under your command to march from their present quarters, by such route and in such divisions as you shall think most convenient (acquainting this office with the route and the day of their arrival at their destined quarters) to Winchester, where they are to relieve His Majesty's Battalion of Militia for the West Riding of the County of York, commanded by Colonel Sir George Saville, in the duty of guarding the French prisoners of war there, and be quartered and remain until further orders.—Given at the War Office this 2nd day of March, 1762. By His Majesty's command."

" To Colonel Vansittart, or Officer commanding His Majesty's Regiment of Militia for the County of Berks at Reading.

" It is His Majesty's pleasure that you cause the quarters of His Majesty's Regiment of Militia for the County of Berks, under your command at Reading, to be enlarged with such adjacent place or places as you shall think proper.—Given at the War Office this 10th day of April, 1762. By His Majesty's command."

During the month of April a Militia Act was passed. Among the chief points in it were that, if preferred instead of ballot, volunteers might be chosen by the parish officers, with the approval of two Deputy Lieutenants. No person under 18 nor over 45, nor articled clerk, apprentice, or poor man with three children born in wedlock, should be compelled to serve. £100 was the penalty for not serving or providing a substitute. The term of service being three years.

"To Officer commanding His Majesty's Regiment of Militia for the County of Berks at Reading.

"It is His Majesty's pleasure that you cause His Majesty's Regiment of Militia for the County of Berks under your command to march from their present quarters in two divisions, according to the routes annexed (acquainting this office with the receipt of this order), to the place of encampment near Winchester, where they are to encamp and remain until further order.—Given to the War Office this 12th day of June, 1762. By His Majesty's command."

"Route for the 1st Division of the Berkshire Militia from Reading to Winchester: Wednesday, June 16th, Basingstoke; Thursday, June 17th, Alresford; Friday, June 18th, Winchester, and encamp. Route for the 2nd Division of the Berkshire Militia from Reading to Winchester: Thursday, June 17th, Basingstoke; Friday, June 18th, Alresford; Saturday, June 19th, Winchester, and encamp."

"To Officer commanding the Berkshire Militia at Reading.

"It is His Majesty's pleasure that, notwithstanding any former order to the contrary, you cause His Majesty's Regiment of Militia for the County of Berks under your command to march from their present quarters, by such routes and in such divisions as you shall think most convenient (acquainting this office with the receipt of this order and the day of their arrival at their destined quarters), so as to assemble at Reading on Monday next, the 21st inst., where they are to be quartered and remain until Wednesday, the 23rd, and Thursday, the 24th inst., when they are to proceed in two divisions (the 2nd day after the 1st), so as to encamp at Winchester on Saturday, the 26th inst., and remain until further order.—Given to the War Office this 17th day of June, 1762. By His Majesty's command."

" To Lieut.-General the Earl of Effingham, or Officer commanding
His Majesty's Berkshire Regiment of Militia at the place of
encampment near Winchester.

" It is His Majesty's pleasure that you cause His Majesty's
Regiment of Militia for the County of Berks under your command
to march from the place of encampment near Winchester, in two
divisions, according to the routes annexed (acquainting this office
with the receipt of this order), to Reading and Oakingham, where
they are to be quartered and remain until further order.—Given to
the War Office this 19th day of October, 1762. By His Majesty's
command."

" Route for the 1st Division of the Berkshire Militia from
Winchester Camp to Oakingham: Monday, October 25th,
Basingstoke; Tuesday, October 26th, Reading; Wednesday,
October 27th, Oakingham and remain. Route for the 2nd
Division of the Berkshire Militia from Winchester Camp to
Reading: Tuesday, October 26th, Basingstoke; Wednesday,
October 27th, Reading and remain."

" To the Officers commanding the following corps of Militia :
Berkshire Regiment, Reading.

" As the time is now drawing near when it may probably be
thought expedient to disembody the Corps of Militia under your
command, I am to signify His Majesty's pleasure that if the
battalion under your command should not in its present distribution happen to be so conveniently quartered as it might be, for
the return of the non-commissioned officers and private men to
the respective divisions of the County of Berkshire from which
they were balloted, you are hereby empowered to march any
companies, parties or detachments, belonging to the battalion
under your command, from their present quarters to any other
place or places within the said county, for the greater convenience
of the said companies, parties or detachments, at the time of their
being disembodied, in doing which you will follow your own
discretion, and be governed by the good of the service and the
convenience of the men.—Given to the War Office this 4th day of
December, 1762. By His Majesty's command."

1762.     Lieut.-Colonel John Dodd, of the Berkshire Militia, was tried at Reading by court martial, on . the complaint of William Mackworth Praed, Esq., one of its captains, for unsoldier-like behaviour, for endeavouring to impede him in his succession to the majority of the said regiment. The finding of the court was as follows :—" The court martial, upon due consideration of the whole matter before them, is of opinion that Lieut.-Colonel Dodd is not guilty of the charge exhibited against him, or any part thereof, and therefore the said court doth acquit him with honour."

Peace was arranged with France in November ; and immediately after the Militia was disembodied, having been under arms about three years.

1763.     April. £150,000 granted for clothing and pay of Militia from March 25th, 1763, and arrangements were made for appointing a time for exercising them annually.

# CHAPTER VIII.

THE Militia Acts were very defective; being subject to frequent delays, many difficulties, and some doubts in the execution. It was therefore thought proper that the following regulations should be observed, viz.: That where the Militia had not been raised, the county have the power to hold meetings on the last Tuesday in May or the last Tuesday in October, in each year. In every county, riding, or place, where the office of Lord Lieutenant was vacant, it should be lawful for His Majesty to appoint three Deputy Lieutenants to execute that office, so far as related to the Acts for raising and training the Militia. No volunteer or substitute was to be admitted and sworn who was not 5ft. 4in. in height. A person being enrolled to serve in the Militia of one county, who engaged to be enrolled to serve in the Militia of another county, forfeited £10, and if not immediately paid was to be committed for any time not exceeding three months.

A Militia man on the march or at the place of exercise, disabled by sickness or otherwise, was to be relieved by the officers of the parish where he happened to be; and the parish officers were to be reimbursed the expenses occasioned thereby out of the county stock, upon producing accounts thereof, allowed by a justice of the peace.

Militia men who, after having joined their corps, deserted during the time of annual exercise, and were not taken till after the expiration of the time of such exercise, should incur the same penalty as Militia men not joining their corps.

A captain, or commanding officer, might put corporals and private men under stoppages not exceeding 6d. a day, and must account to them for such stoppages before they were dismissed from annual exercise.

A drummer negligent in his duty, or disobedient to the orders of the adjutant or other superior officer, was to forfeit a sum not exceeding 40s., and if not immediately paid, the captain of the company was to stop the pay of such drummer to pay the penalty, which penalty was to be applied as part of the common stock of the regiment or battalion.

Where the Militia should not be raised for any county, within which any city should not be rated to the county rate, the payment of £5 per man should be apportioned between such county or city as the respective quotas paid to the land-tax bear to each other; and the sums so apportioned should be paid out of the poor rate collected in such city by the churchwardens and overseers of the poor to the Treasurer of the county, to be by him paid to the Receiver-General, together with the proportion of the said sum of £5 to be paid by such county. The same method was to be followed in such cities as were counties in themselves; and where a town lies in two counties they were to contribute their quota, in lieu of raising the Militia for that county in which their church stands, and the deficiencies of the other county rates were to be made up by the county in general. Similar Acts relating to the Militia are to be found in the *Universal Magazine* for May, 1762, and April, 1763.

1766. A board of general officers was held at the Horse Guards, under the presidency of Lord Viscount Ligonier, to take under consideration and establish a rule as to the future purchasing of commissions in the land service, and ascertaining the purchase-money to be paid. In the time coming all brokers of commissions were to be laid aside, no subaltern or other officer was to be appointed unless the consent and approbation of the colonel or commanding officer of the regiment had first been obtained. The last board of this nature was held so long ago as 1725.

For some years after the disembodiment, training of the Berkshire Militia took place annually at Reading or Newbury, which is curious, as Abingdon was the county town : nor was Wallingford ever patronised.

This was the year of great distress, terminating in riots all over England. The seasons seemed all out of order. In July terrible rains caused floods ; at Maidenhead and other places in Berkshire, the Thames rose and much damaged the hay.

Newbury seems to have been conspicuous for riots in August. The people rose in riot in the Market-place. They ripped up the sacks, scattered the wheat, took butter, meat and bacon out of the shops and threw it into the streets. The bakers were frightened and reduced the price of their loaves 2d. a peck, and promised another reduction for the following week. At Shaw Mill they threw the flour into the river, broke the windows of the house, and then went on and damaged several other mills. But the rioters did not get off scot free. A special Commission was held at Reading, three rioters were sentenced to death, and stringent measures were taken to try and suppress the riots, and soldiers were called out. While Acts of Parliament and private liberality tried every means of relieving the distress of the poor.

1767.  An order in Council was published in the *London Gazette*, requiring lieutenants of counties where the Militia have been embodied, to make out lists of the officers, to prevent their being nominated for Sheriffs during the time of their employment in the service.

1768.  May 2nd. The Berkshire Regiment was ordered to assemble at Newbury.

On Wednesday, May 25th, the Regiment of Militia for this county, quartered at Newbury for their annual exercise, were reviewed by their colonel. They went through their firing and manœuvres entirely to the satisfaction of a large number of spectators, amongst whom were many officers of the Army, who allowed them to be as well-disciplined as any regiment in His Majesty's service. They were afterwards disembodied.

A week later John, *alias* Peter Castle, was committed to Reading Gaol, by C. Saxton, Esq., for not joining the Regiment of Berkshire Militia at their last rendezvous at Newbury. He was ordered to remain in gaol six months, or until he could pay a fine of £20. One or two cases occurred every year, and the penalties were rigidly enforced. So that desertion, after receiving the bounty money, was not done with impunity.

December. Serious rains fell, causing floods. The Kennet and Loddon overflowed, Burghfield Bridge was washed away, and part of Twyford Bridge. The whole country was like a sea.

1769.     October 4th. The Militia came out for 28 days training at the Forbury, Reading.

On the anniversary of His Majesty's Accession to the Crown in November, to celebrate the event, the officers of the Berkshire Regiment then on duty gave a ball in the Town Hall, Reading, on so loyal an occasion, was rendered exceedingly brilliant by the appearance of ladies from all parts of the neighbourhood, as well as being honoured by the presence of the Right Honourable the Lord Chancellor, and many other persons of distinction. On Wednesday the said regiment, having performed the annual 28 days' exercise, the men were disembodied and returned to their respective habitations.

1770.     May 7th. The training took place at Newbury, under Colonel Vansittart. They dispersed, after

28 days, on June 5th. One or two privates were arrested and imprisoned for having failed to appear for training in 1767.

Correspondence *re* New Colours:

"To Lord Viscount Weymouth.

" Office of Ordnance, May 22nd, 1770.

" My Lord,—Having this day received your Lordship's letter, dated the 11th of October, 1769, directing a Sett of new Colours to be issued for the use of the Berkshire Militia, we beg leave to acquaint your Lordship, that it is impossible to get the said Colours made in time for the annual exercise this year, which will expire the 2nd of next month; however, it is proper we should acquaint your Lordship, that a sett of Colours (as we are informed) generally serves much longer than twelve years for His Majesty's Regiments, and that the Berkshire Militia were supplied with new Colours, in 1758.—We are, my Lord, your most obedient and most humble servants, CHARLES FREDERICK, A. WILKINSON, W. R. EARLE, CHAS. COCKS."

" To Lieutenant-General and other principal Officers of the Ordnance.

" St. James's, May 29th, 1770.

" Gentlemen,—Yesterday I received the favor of your letter, dated the 22nd inst., in which you inform me that mine of the 11th of October, 1769, in which I signified his Majesty's pleasure that a new set of Colours should be issued for the use of the Berkshire Militia, was just delivered at your Board, that it is impossible to get the said Colours made in time for the annual exercise this year; that a set of Colours (as you are informed) generally serves much longer than twelve years for His Majesty's Regiments; and that the Berkshire Militia were supplied with new Colours in 1758. As to the delay of my letter, I can only say that it was put into the hands of the person who applied for the Colours above mentioned; and as to the propriety of delivering them, I must leave that to your determination, having sent a copy of your letter to Lord Vere, the Lord Lieutenant of the county, who will apply to you for the proper certificate conformable to my letter of the 15th inst.—I am, &c., WEYMOUTH."

*From a portrait at Purley Hall.*

Major JOHN WILDER.
1769.

October 30th. A general meeting was held by the Deputy Lieutenants of Berks. Present: Arthur Vansittart, Esq., John Walter, Esq., and Peter Floyer, Esq. This list gives the division and sub-divisions of the county where meetings of the lieutenancy were held:

| SUB-DIVISION. | DATE. | PLACE. |
|---|---|---|
| Abingdon, containing the Hundreds of Hormer, Oar and Moreton, and the towns of Abingdon and Wallingford - | 1st Meeting, Monday, December 31st - - <br> 2nd, first Monday in March - - - - <br> 3rd, first Monday in June - - - - - <br> 4th, first Monday in September - - - | At "The Crown and Thistle," Abingdon. |
| Farringdon, containing the Hundreds of Farringdon, Sanfield, and Shrivenham - - - - - | Tuesday, January 1st - <br> 1st Tuesday in March <br> " June - <br> ,, Sept. - | At "The Bear," Farringdon. |
| Forest, containing the Hundreds of Charlton, Sonning, and Wargrave - - - | Tuesday, January 1st - <br> March <br> June <br> Sept. | At "The Rose," Wokingham. |
| Maidenhead, containing the Hundreds of Beynhurst, Bray, Cookham, Rufflesmere, and the towns of New Windsor and the Castle - - | Wednesday, January 2nd - - - - - <br> March <br> June <br> September | At "The Bear," Maidenhead. |
| Reading, containing the Hundreds of Reading, Theale, and the town of Reading - - - - | Saturday, January 5th <br> March <br> June <br> September | At "The Three Tuns," Reading |

H

| Newbury, containing the Hundreds of Compton, Faircross, Kintbury, Eagle, and the town of Newbury - | Thursday, January 3rd - - - - - - March June September | At "The Globe," Newbury. |
| Wantage, containing the Hundreds of Wantage and Lambourne - - - - | Saturday, January 5th March June September | At "King Alfred's Head," Wantage. |

(Signed)    WILLIAM TYRRELL,

Clerk to the General Meeting.

1771.    General orders were issued that the Militias should, like the Regulars, have a light company and a band to every battalion. While embodied the Militia Regiments were, in every way, made equal to permanent regiments; indeed, in many ways, they were superior to the latter, as their officers did not spare expense to make their men smart and efficient.

1772.    In April a Board of General Officers, who sat by Royal mandate, at the Horse Guards last Tuesday, on the reference "whether the rank of Major in the Army should be totally abolished or not?" Decided, "That the rank should remain as at present."

The Berkshire Militia assembled for training in the Market Place, Newbury, on May 4th, as arranged by the Deputy Lieutenant of the County, at a Meeting held in the Reading Town Hall, two months previously.

"War Office, May 26th, 1772.

"The King has been pleased to direct that, for the future, the Captain-Lieutenants of the cavalry and marching regiments shall have rank, as well in the Army as in their respective regiments, as Captains; that the present Captain-Lieutenants shall take the said rank from this day, and all future Captain-Lieutenants from the date of their respective commission.—BARRINGTON."

1773.
While the Militia were out for training at Reading, one of the privates, John Gibbs by name, died and was buried with military honours in the churchyard of St. Mary's Church.

1774.
The County of Nottingham was fined, at the General Quarter Sessions for Nottinghamshire, upwards of £2,000 for not raising the Militia the previous year. An order was made to levy the money. This shews that in some parts of England there *was* difficulty in finding soldiers; certainly Berkshire had little difficulty in the matter compared to other counties.

While the Berks Militia were out for their training at Newbury, a catastrophe happened. The old Mansion of Benham, about two miles from the town, caught fire. Immediately it was perceived at Newbury, the drums of the Militia beat to arms. The men were marched up to the place with all speed by their officers; and by their activity and regularity a great part of the valuable furniture was saved. Had there been a greater number of buckets, it is thought some part of the house might have been saved, as there was a sufficient cordon of men to extend to the river, though it is some distance from the house. The kitchen, stables, and some offices were saved. An officer's guard was mounted for the security of the furniture. Lord and Lady Craven were away in London at the time. The inhabitants of Newbury were jealous of the credit being given to the Militia for their efforts to quell the fire. Though not wishing to detract from the credit due to the regiment, the townspeople were anxious to share in the praise and honour themselves.

1775.
At his quarters at the " Barley Mow " Inn, London Street, Reading, Sergeant Pittman, of the Berks Militia, shot himself through the head with a horse pistol. He had served many years in Albemarle's Dragoons, was present at the Battle of Dettingen, and several other

H 2

actions. When the Militia was embodied in 1757, he obtained his discharge from the Regulars, and was appointed Sergeant in the Militia. His good conduct had gained the approbation of his officers and the respect of his comrades. No cause except lunacy could be found to account for the suicide, and this verdict was given at the inquest.

May 23rd. The officers of the Militia, with the Band of the Regiment, went for an excursion by water to Hardwick. The banks of the Thames on either side were lined with villagers to see them pass, who were delighted with the fine appearance of the boats and with the music. An elegant cold collation was provided and served at Straw Hall, an agreeable villa belonging to Philip Powys, Esq., who had lately married ; and the afternoon was spent in harmony and festivity.

The anniversary of the Restoration of Charles II. was celebrated in Reading by the ringing of Church bells. The military were drawn up in the Market Place at 12 o'clock and fired three volleys, and the day was marked with demonstrations of joy.

The Militia were reviewed on Bulmarsh Heath, by Colonel Vansittart; performing their exercises with great steadiness, and going through their evolutions with an alertness and regularity that, in the opinion of many of the officers of the Army present, would have done credit to any regiment in His Majesty's service. The following day they were dismissed, the 28 days allowed for training having been completed ; but it was rumoured they might soon expect to be again called out.

1777.      Although unconnected with our Militia, I must here mention an episode in military history which strikes me as amusing. Lord George Lennox's Regiment was quartered in Winchester. That autumn, workmen being scarce and harvest in full swing, his Lordship gave his whole regiment leave to work for the farmers, who employed them at 10s. per week and their board. The soldiers went reaping with great cheerfulness every morning, with drums beating, music playing, and colours flying. It appears that the commanding officer had the right to employ his regiment as he thought fit, and soldiers were sometimes put to duties other than mere drill and exercise.

1778. Orders were issued in March to the Lords-Lieutenants of the counties, from the Secretary of State's office, ordering the Militia of each county to be immediately embodied.

The following is the authentic account of the summer encampment of the English troops; but, unfortunately, no mention is made as to which Militia regiments were in the camps, except in the case of Warley Camp. The Berkshire Militia were among those at Coxheath Camp, the chief rendezvous of the troops, and the following year were sent to Warley :—At Salisbury : 1st, 2nd, 3rd, and 6th Dragoon Guards. St. Edmundbury: 3rd, 4th, 7th, and 10th Dragoons. Coxheath, Kent: 1st Battalion of Royals; 2nd, 14th, 18th, 59th, and 65th Regiments of Foot, 1st Regiment of Dragoons, and twelve Regiments of Militia. Warley Common, Essex : 6th, 25th, and 69th Regiments of Foot ; and six Regiments of Militia, viz., the Somerset, Wilts, Kent, Carmarthen, Glamorgan, and Pembroke. Winchester : 30th Regiment of Foot, and six Regiments of Militia. Plymouth : Three Regiments of Militia. Portsmouth : Two Regiments of Militia. Dover: One Regiment of Militia.

Marching Orders :

" To Officer commanding the Berkshire Militia.

"It is His Majesty's pleasure that you cause the Berkshire Regiment of Militia under your command to march from their present quarters, by the shortest and most convenient route, on Saturday, the 23rd inst., to Henley, Oakingham, and Great Marlow, where they are to be quartered and remain until further order. You will also acquaint this office with the receipt and execution of this order.—Given at the War Office this 18th day of May, 1778."

"To Officer commanding the Berkshire Regiment of Militia at
    Henley.

"It is His Majesty's pleasure that you cause the Berkshire
Regiment of Militia under your command to march from their
present quarters in two divisions, according to the routes annexed,
to Coxheath, where they are to encamp and remain until further
order. You will also acquaint this office with the receipt and
execution of this order.—Given at the War Office this 1st day of
June, 1778."

"Route for the 1st Division of the Berkshire Regiment from
Henley to Coxheath camp : Saturday, June 6th, Windsor, Datchet,
Slough, and Salthill ; Sunday, June 7th, halt ; Monday, June 8th,
Kingston ; Tuesday, June 9th, Bromley ; Wednesday, June 10th,
Sevenoaks ; Thursday, June 11th, halt ; Friday, June 12th,
Wrotham, Wrotham Heath, Trottscliff, Opham, and Mallings ;
Saturday, June 13th, encamp on Coxheath and remain. Route
for the 2nd Division of the Berkshire Militia from Henley to
Coxheath : Saturday, June 6th, Beaconsfield ; Sunday, June 7th,
halt ; Monday, June 8th, Acton, Ealing, and the Old Hats ;
Tuesday, June 9th, Greenwich, Blackheath, and Deptford ; Wed-
nesday, June 10th, Dartford and Crayford ; Thursday, June 11th,
halt ; Friday, June 12th, Rochester, Chatham, Stroud, and
Finsbury ; Saturday, June 13th, encamp on Coxheath and remain."

"To Officer commanding the companies of the Berkshire Militia
    at Oakingham.

"It is His Majesty's pleasure that you cause the two companies
of the Berkshire Regiment of Militia under your command at
Oakingham to march from thence on Friday next, the 5th inst., to
the places mentioned in the margin hereof, where they are to be
quartered until the next day, and then follow the order of this date
for the march of the regiment to the camp on Coxheath.—Given
at the War Office this 1st day of June, 1778."

"To Lieut.-General Keppel, or Officer commanding the companies
    of the Berkshire Militia at Coxheath camp.

"It is His Majesty's pleasure that you cause the Berkshire
Regiment of Militia under your command to march from the place

of their encampment in two divisions, according to the routes annexed, to Reading, where they are to be quartered and remain until further order.—Given at the War Office this 5th day of November, 1778."

"Route for the march of the 1st Division, consisting of five companies of the Berkshire Regiment of Militia, from Coxheath camp to Reading: Monday, November 9th, Wrotham, Wrotham Heath, East and West Mallings, Offham, Ightam, Otford, Sevenoaks, Seal, and Riverhead; Tuesday, November 10th, Bromley and Croydon; Wednesday, November 11th, Fulham, Putney, and Handsworth; Thursday, November 12th, halt; Friday, November 13th, Staines, Egham, and Egham Hithe; Saturday, November 14th, Oakingham; Sunday, November 15th, halt; Monday, November 16th, Reading and remain. Route for the march of the 2nd Division, consisting of four companies of the Berkshire Regiment of Militia, from Coxheath Camp to Reading: Monday, November 9th, Wrotham, Wrotham Heath, East and West Mallings, Offham, Ightam, Otford, Sevenoaks, Seal, and Riverhead; Tuesday, November 10th, Bromley and Croydon; Wednesday, November 11th, Fulham, Putney, and Handsworth; Thursday, November 12th, halt; Friday, November 13th, halt; Saturday, November 14th, Staines, Egham, and Egham Hithe; Sunday, November 15th, halt; Monday, November 16th, Oakingham; Tuesday, November 17th, Reading and remain.

"BARRINGTON."

1779.    December. While the Militia were back again, quartered in Reading, a wedding took place at St. Mary's Church, between William Lambe and Elizabeth Reille, by license and consent of parent.

## BERKSHIRE MILITIA.

LIST OF THE OFFICERS FOR THE YEAR 1779.

| Rank. | Name. |
|---|---|
| Colonel - - - - - - | Arthur Vansittart. |
| Lieut.-Colonel - - - - | Charles Saxton. |
| Major - - - - - - - | John Walter. |
| Captains - - - - - - | Edmund Seymour. |
| | Pennston Powney. |
| | Henry James Pye. |
| | Edward Loveden Loveden. |
| | John Charles Price. |
| | George Elwes. |
| Captain-Lieutenant - - | Edward Sheppard. |
| Lieutenants - - - - - | Henry Evans. |
| | John Fortescue Acland. |
| | Walter Pye. |
| | William Sladden. |
| | John Blagrave. |
| | Thomas Groves. |
| | Richard Aldworth Neville. |
| | John Stephenson. |
| | Joseph Blagrave. |
| Ensigns - - - - - - | Robert Parker. |
| | William Nathaniel French. |
| | Thomas Velley. |
| | John Wallis. |
| | Philip Gill. |
| | Joseph Hervey Bellas. |
| | Charles George Starck. |
| Adjutant - - - - - - | Henry Evans. |
| Quartermaster - - - - | William Sladden. |
| Surgeon - - - - - - | John Fortescue Acland. |

| | |
|---|---|
| Uniform - - - - - - | Red, faced with Light Blue. |
| Agents - - - - - - | Messrs. Cox & Mair, Craig's Court |

" To the Officer commanding the Berkshire Militia at Reading.

" It is His Majesty's pleasure that you cause two companies of the Berkshire Regiment of Militia under your command to march from their present quarters, by the shortest and most convenient route, to the places mentioned in the margin hereof, where they are to be quartered and remain until further order. You will also acquaint this office with the receipt and execution of this order.—Given at the War Office this 20th day of February, 1779. By His Majesty's command,

Companies:
Banbury ... 1
Woodstock 1
_
2

" C. JENKINSON.

" To Colonel Vansittart, or Officer commanding the Berkshire Militia at Reading ; like order of the same date to Lieut.-General Johnstone, Oxford.

" It is His Majesty's pleasure that you cause such parties of the Berkshire Regiment of Militia under your command as shall be found expedient, on account of the approaching expiration of the term of service of several Militiamen belonging to the said corps, to march to and from the respective quarters of the said regiment at such times and in such detachments as may be judged necessary, and be quartered as occasion shall require. And it is His Majesty's further pleasure that you cause the men whose time is nearly expired to march to such places in the County of Berks as shall be thought proper, and the new ballotted men to join the regiment.—Given at the War Office this 1st day of April, 1779. By His Majesty's command, M. LEWIS."

" To Officer commanding the Berkshire Militia at Reading. Sent to Lieut.-General Johnson, South Audley Street.

" It is His Majesty's pleasure that you cause a party, consisting of one subaltern and twenty private men, to be made from the Berkshire Militia under your command ; and receive from the Berkshire Gaol, at Reading, several impressed men for His Majesty's service, and escort them by the shortest and most convenient route to Slough, where they are to be delivered to such other party as shall be appointed to receive them, and after the performance of this service, the said party are to return to their present quarters.—Given at the War Office this 26th day of April, 1779."

" To Officer commanding the Berkshire Militia at Woodstock.

" It is His Majesty's pleasure that you cause a detachment, consisting of one sergeant, one corporal, and sixteen private men, to be made from the companies of the Berkshire Militia at Woodstock, and march on Monday next, May 3rd, to Oxford, where they are to take charge of the impressed men there, and safely escort them by the shortest and most convenient route to High Wycombe, and deliver them over to a like party of the Bucks Militia and be quartered, and return the next day and join the companies to which they belong, and remain until further order. You will also acquaint this office with the receipt and execution of this order.—Given at the War Office this 27th day of April, 1779."

" To Officer commanding the companies of the Berkshire Militia at Reading.

" It is His Majesty's pleasure that you cause the seven companies of the Berkshire Militia under your command at Reading to march from thence on Monday, the 24th inst., by the shortest and most convenient routes to the places mentioned in the margin hereof, where they are to remain till Wednesday, the 26th inst., and then return to their present quarters and remain until further order.—Given at the War Office, this 19th day of May, 1779. By His Majesty's command,

Henley, Nettlebed, Wallingford, and Bensington.

"C. JENKINSON."

" To Officer commanding the companies of the Berkshire Militia at Reading.

" It is His Majesty's pleasure that you cause the companies of the Berkshire Militia at Woodstock and Banbury to march from their present quarters by the southern and midland counties route, so as to join their regiment at Reading on Saturday, June 5th next and be quartered and remain until further order. You will also acquaint this office with the receipt and execution of this order.—

Given at the War Office this 29th day of May, 1779. By His Majesty's command, in the absence of the Secretary at War,

"M. LEWIS."

" To Officer commanding the Berkshire Militia at Reading.

" It is His Majesty's pleasure that you cause the Berkshire Regiment of Militia under your command to march from their present quarters, according to the route annexed, to Adarley Common, and encamp and remain until further order. And it is His Majesty's further pleasure that you acquaint this office with the receipt and execution of this order.—Given at the War Office this 1st day of June, 1779."

" Route for detachment from Reading to Adarley camp : Monday, June 7th, Henley, Maidenhead, Maidenhead Bridge, and Great Marlow ; Tuesday, June 8th, Slough, Salthill, Colnbrook, Longford Bridge, Cranford Bridge, and Hounslow ; Wednesday, June 9th, Waltham Green, Hammersmith, Turnham Green, Kensington, and Kensington Gravel Pits ; Thursday, June 10th halt ; Friday, June 11th, Rumford, Hare Street, and Ilford ; Saturday, June 12th, encamp at Adarley Common."

" To Officer commanding the Berkshire Regiment of Militia at Rumford.

" It is His Majesty's pleasure that, notwithstanding any former órder, you cause the Berkshire Regiment of Militia under your command to be quartered at Rumford, Ilford, and Hare Street, until they can proceed to the place of encampment and be quartered, and remain until further order. You will also acquaint this office with the receipt and execution of this order.—Given at the War Office this 10th day of June, 1779. By His Majesty's command, C. JENKINSON."

" To Officer commanding the Berks Militia at Woodstock.

" The Right Honourable the Lords Commissioners of the Admiralty having requested that orders may be given for the escort of one thousand Spanish prisoners of war from Portsmouth to Shrewsbury in four divisions, it is His Majesty's pleasure that on the arrival of each division of the said prisoners at Woodstock, under escort of the Berkshire Militia, you cause four successive detachments (each to consist of a captain commandant) to be made from the Bucks Militia under your command at Woodstock and receive each division of the said prisoners, and be assisting in safely escorting them to Chipping Norton, where they are to be delivered to the officer commanding the companies of the said Militia at that place, who is hereby required to receive the said prisoners and furnish like escorts for conducting them to Stratford-upon-Avon, where they are to be delivered to the officer commanding the troops of the 11th and 22nd Regiments of Dragoons, who has orders to receive them ; the detachments, after the performance of this service, are to return and join their corps. You will also acquaint this office with the receipt and execution of this order.—Given at the War Office this 13th day of March, 1780."

" To Officer commanding the Berkshire Militia at Reading.

" It is His Majesty's pleasure that you cause five companies of the Berkshire Militia under your command to march from their present quarters on Tuesday, the 30th inst., to Basingstoke, from whence they are to proceed the next day by the southern and midland counties route to Winchester (17¾ miles) and relieve the companies of the Stafford Militia in the duty on the prisoners of war, and be quartered and remain until further order. You will also acquaint this office with the receipt and execution of this order.—Given at the War Office this 24th day of May, 1780."

" To Officer commanding the four companies of the Berkshire Militia at Reading.

" It is His Majesty's pleasure that you cause the four companies of the Berkshire Militia under your command at Reading to march from thence, according to the route annexed, to Hilsea Barracks, and be quartered and remain until further order. You will also acquaint this office with the receipt and execution of this order.— Given to the War Office this 29th day of May, 1780."

" Route for detachment from Reading to Hilsea Barracks : Saturday, June 10th, Basingstoke (17 miles) ; Sunday, June 11th, halt ; Monday, June 12th, Alton, Chewton, and Farringdon; Tuesday, June 13th, Petersfield ; Wednesday, June 14th, Hilsea Barracks and remain."

" To the Honourable Lieut.-General Monckton, or Officer commanding the forces at Portsmouth.

" It is His Majesty's pleasure that you cause the Berkshire Militia under your command to march from their present quarters in two divisions, according to the routes annexed, to the places mentioned in the margin hereof, where they are to be quartered and remain until further order. You will also acquaint this office with the receipt and execution of this order.—Given at the War Office this 13th day of October, 1780. By His Majesty's command, in the absence of the Secretary at War,

| Companies : | |
| --- | --- |
| Banbury and Newthorpe ... ... | 2 |
| Burford ... .. | 1 |
| Deddington, Adderbury, & Bloxham | 1 |
| Witney & Eynsham | 1 |
| Chipping Norton & Chapel House ... | 2 |
| Bicester ... ... | 1 |
| Islip & Blechingdon | 1 |
| | 9 |

" M. LEWIS."

"Route for the 1st Division of the Berkshire Regiment of Militia, consisting of five companies, from Hilsea Barracks to Banbury (nearly 60 miles across country), &c. : Tuesday, October 17th, Petersfield ; Wednesday, October 18th, Alton, Chewton, and Farringdon: Thursday, October 19th, halt ; Friday, October 20th, Basingstoke ; Saturday, October 21st, Reading ; Sunday, October 22nd, halt ; Monday, October 23rd, Wallingford, Cromarch, and Bensington ; Tuesday, October 24th, Abingdon ; Wednesday, October 25th, Whitney (2), Eynsham, where it is to remain, and Woodstock (3) ; Thursday, October 26th, halt ; Friday, October 27th, Burford from Whitney (1), Banbury, Newthorpe (2), Doddington, Adderbury, Bloxham (1), and remain. Route for the 2nd Division of the Berkshire Regiment of Militia : Wednesday, October 18th, Petersfield ; Thursday, October 19th, halt ; Friday, October 20th, Alton, Chewton, and Farringdon ; Saturday, October 21st, Basingstoke ; Sunday, October 22nd, halt ; Monday, October 23rd, Reading ; Tuesday, October 24th, Wallingford, Cromarch, and Bensington ; Wednesday, October 25th, Abingdon ; Thursday, October 26th, halt ; Friday, October 27th, Oxford ; Saturday, October 28th, Chipping Norton, Chapel House, Belchingdon (2), &c., and remain."

"To Officer commanding the company of the Berkshire Militia at Bicester.

" It is His Majesty's pleasure that you cause the company of the Berkshire Militia under your command at Bicester to march on Monday, the 13th inst., by the southern and midland counties route, to Old and New Woodstocks, where they are to be quartered and remain until Thursday, the 16th inst.; and then return to their present quarters and remain until further order.—Given at the War Office this 7th day of November, 1780."

" It is His Majesty's pleasure that you cause the quarters of the company of the Berkshire Militia under your command at Islip, &c., to be enlarged with Kidlington and Kirklington.—Given at the War Office this 10th day of November, 1780."

" To Officer commanding the two companies of the Berkshire Militia at Chipping Norton, &c.

" It is His Majesty's pleasure that you cause one of the two companies of the Berkshire Militia under your command at Chipping Norton and Chapel House to march from thence on Friday next, the 29th inst., by the southern and midland counties route, to Charlbury, Enton Shipton-under-Whichwood, and such other public houses in the neighbourhood of Charlbury as you shall judge proper.—Given at the War Office this 26th day of December, 1780."

" To Officer commanding the Berkshire Militia at Banbury, &c.

Companies :

| | |
|---|---|
| Barnet, Hadley, Kitt's End, Ridge Mims, Potter's Bar, & Northall | 4 |
| Whetstone ... ... | 1 |
| Hampstead, Highgate, Hornsey, & that part of the Parish of St. Pancras within the hamlet of Highgate... | 3 |
| Stanmore, Edgware and Bushey ... ... | 1 |
| | — |
| | 9 |

" It is His Majesty's pleasure that you cause the Berkshire Militia under your command to march from their present quarters in two divisions, according to the routes annexed, to the places mentioned in the margin hereof, where they are to be quartered and remain until further order. You will also acquaint this office with the receipt and execution of this order. —Given at the War Office this 6th day of April, 1781."

"Route for the 1st Division of the Berkshire Regiment of Militia, consisting of five companies, from Deddington, Bicester, Islip, Witney, and Burford : Monday, April 9th, Oxford, Islip, and Bicester ; Tuesday, April 10th, Thame, Titsworth, and Wheatley ; Wednesday, April 11th, High and West Wycombe ; Thursday, April 12th, halt ; Friday, April 13th, Walford and Rickmansworth ; Saturday, April 14th, Barnet, Hadley, Kitt's End, Ridge, Mims, Potter's Bar, Northall (4) and remain, Whetstone (1) and remain. Route for the 2nd Division of the Berks Regiment of Militia, consisting of four companies, from Banbury, Chipping Norton, and Charlbury : Monday, April 9th, Old and New Woodstock ; Tuesday, April 10th, Oxford ; Wednesday, April 11th, Thame, Titsworth, and Wheatley ; Thursday, April 12th, halt ; Friday, April 13th, High and West Wycombe ; Saturday, April 14th, Uxbridge and Hillingdon ; Sunday, April 15th, halt ; Monday, April 16th, Hampstead, Highgate, Hornsey, and that part of the Parish of St. Pancras within the hamlet of Highgate (3), and remain, Stanmore, Edgware and Bushey (1), and remain." London to Deddington is 69 miles.

Three companies of the Berkshire Militia at Hampstead and Highgate were ordered to march on Saturday, the 28th inst., to Barking, Ilford, Bow, Bromley and Stratford, and remain until further orders."

To Officer commanding the Berkshire Militia at Hampstead.

" It is His Majesty's pleasure that you cause one of the companies of the Berkshire Militia under your command at Barnet to march from thence to Edgware, and the company at present at that place to proceed to Barnet, where they are to be quartered and remain until further order.—Given at the War Office this 17th day of April, 1781. By His Majesty's command,

"C. JENKINSON."

" To Officer commanding the Berkshire Militia at Barnet. [By express.]

" It is His Majesty's pleasure that you cause the Quartermaster and camp colour men of the Berkshire Militia under your command to march immediately from their present quarters, by the southern and midland counties route, to Lenham, on Monday next, the 7th inst., where they are to be quartered and follow such directions as they shall receive from the assistant to the Quarter-master-General, on his arrival at the place. You will also acquaint this office with the receipt and execution of this order.—Given at the War Office this 2nd day of May, 1781. By His Majesty's command, in the absence of the Secretary at War, M. LEWIS."

" To Officer commanding the Berks Militia at Barnet, &c.

Companies:
Maidstone ... 5
Sevenoaks, Seal,
and Riverhead 2
Wrotham, Igh-
tam, Offham,
and Mallings 2
—
9

"It is His Majesty's pleasure that you cause the Berks Militia under your command to march from their present quarters (leaving the detachment at Paddington) in two divisions, according to the routes annexed, to the places mentioned in the margin hereof, where they are to be quartered and remain until further order. You will also acquaint this office with the receipt and execution of this order.—Given at the War Office this 3rd day of May, 1781." London to Maidstone is about 34 miles.

" Route for the 1st Division of the Berkshire Regiment of Militia, consisting of five companies, from Barnet to Maidstone, &c. : Monday, May 7th, Lambeth, Vauxhall, and Newington ; Tuesday, May 8th, Bromley and Beckenham ; Wednesday, May 9th, Sevenoaks, Seal, and Riverhead ; Thursday, May 10th, halt ; Friday, May 11th, Maidstone and remain. Route for the 2nd Division of the Berkshire Regiment of Militia, consisting of four companies, from Barnet to Bow, Edgware, &c. : Tuesday, May 8th, Lambeth, Vauxhall, and Newington ; Wednesday, May 9th, Bromley and Beckenham ; Thursday, May 10th, halt ; Friday, May 11th, Sevenoaks, Seal, Riverhead and remain ; Saturday, May 12th, Wrotham, Ightam, Offham, Mallings (2) and remain."

To Officer commanding the party of the Berkshire Militia at Paddington. To be sent to Captain Green, No. 3, Panten (Paddington ?) Square.

"It is His Majesty's pleasure that you cause the party of the Berkshire Militia under your command at Paddington to march from thence, by the southern and midland counties route, and join the regiment at Maidstone, and be quartered and remain. —Given at the War Office this 7th day of May, 1781."

" To Officer commanding the Berkshire Militia at Maidstone.

" It is His Majesty's pleasure that you cause a sergeant and ten men of the Berkshire Militia under your command to march immediately to Charing and Egerton, where they are to be quartered, and assist in clearing the ground upon which the regiment is to encamp.—Given at the War Office this 31st day of May, 1781."

" To Officer commanding the Berkshire Militia at Maidstone.

" It is His Majesty's pleasure that you cause the Berkshire Militia under your command to march from their present quarters on Wednesday, June 6th next, and be quartered at the places mentioned in the margin, from whence they are to proceed on Thursday, the 7th, and encamp on Lenham Heath.—Given at the War Office this 31st day of May, 1781. By His Majesty's command, C. JENKINSON."

" To Lieut.-General Fraser, commanding the Berkshire Regiment of Militia at Lenham Camp.

" It is His Majesty's pleasure that you cause the Berkshire Militia under your command to march from the place of encampment, according to the route annexed, to the places mentioned, viz. : Sevenoaks, Seal, and Riverhead, 2 companies ; Tunbridge and Hadlow, 1 ; Tunbridge Wells, with its environs, 3 ; Lamberhurst, Goudhurst, and Horsemunden, 1 ; Cranbrooke, Milkhouse Street. and Hawkhurst, 1 ; Westerham, Brasted and Sunbridge, 1, where they are to be quartered and remain until further notice.— Given at the War Office this 20th day of October, 1781. By His Majesty's command, in the absence of the Secretary at War,

" M. LEWIS."

" Route for the Berkshire Militia from Lenham Camp to Sevenoaks, &c. : Wednesday, October 31st, Maidstone (7), Smarden (2), Hedcorn, and Staplehurst ; Thursday, November 1st, halt ; Friday, November 2nd, Sevenoaks (3), Seal and Riverhead, where two are to remain ; Saturday, November 3rd, Tunbridge (1) and

I

Hadlow, and remain; Sunday, November 4th, Tunbridge Wells (3) and its environs, and remain; Monday, November 5th, Lamberhurst (1), Goudhurst and Horsmonden, and remain; Tuesday, November 6th, Cranbrook (1), Milkhouse Street and Hawkhurst, and remain; Wednesday, November 7th, Westerham (1), Brasted and Sundridge, and remain."

" To Officer commanding the Berkshire Regiment of Militia at Lenham Camp.

" It is His Majesty's pleasure that you cause the company of the Berkshire Militia destined for Cranbrook (notwithstanding any former order) to halt at Smarden, &c., on Friday, November 2nd., and on Saturday, the 3rd, proceed to Cranbrook, Milkhouse Street and Hawkhurst, and be quartered and remain until further order. —Given at the War Office this 23rd day of October, 1781."

" To Officer commanding the Berkshire Militia at Tunbridge Wells.

" It is His Majesty's pleasure that you cause the quarters of the companies of the Berkshire Militia under your command at Tunbridge Wells, to be enlarged with Wadhurst.—Given at the War Office this 17th day of November, 1781."

" To Officer commanding the companies of the Berkshire Militia at Tunbridge Wells.

" It is His Majesty's pleasure that you cause the quarters of the companies of the Berkshire Militia under your command at Tunbridge Wells, to be enlarged with the Parish of Brenchley. — Given at the War Office this 23rd day of November, 1781. By His Majesty's command, C. JENKINSON."

Orders were given for a general Muster and return of all the Militias, and all vacancies were commanded to be at once filled up.

1780-82. The Muster Rolls give the names of officers, men balloted for, men absent and by whose leave.

1781. The establishment of the Colonel's Company was: One Colonel, one Lieutenant, one Ensign, three Sergeants, three Corporals, two Drummers, and fifty-seven Privates, but only forty-two were present ; the Lieut.-Colonel, the Lieutenant, one Sergeant, one Corporal, and five Privates were absent. Arthur Vansittart, Colonel ; Edward Sheppard, Capt.-Lieutenant; Ashburnham Newman Toll, Ensign. There were an Adjutant, Quartermaster, two Surgeons, three Sergeants, three Corporals, two Drummers, and fifty-seven Privates.

Lieut.-Colonel Clement Sexton's Company : Joseph Butler, Lieutenant ; John Cartwright Blake, Ensign.

June 25th. Captain Groves' Company : Philip Gill, Lieutenant ; Francis Hawes, Ensign. When the return was sent in they were at Southampton.

Captain William Sladden's Company : Henry Evans, Lieut.; John Fonblanque and Osborne Tylden, Ensigns.

Captain Edmund Seymour's Company : Philip Gill, Lieutenant ; Francis Hawes, Ensign.

Most of the companies were under strength. About fifty-six privates seemed the usual strength, two drummers to each company. I have not transcribed the lists of each company. The lists give the absentees, who balloted for, and the names of the men, &c.

June 25th to December 24th. Capt. Penyston Powney's Company : John Wallis, Lieutenant ; William Cleveland, Ensign ; and forty-two Privates.

Captain John Charles Price's Company : James Baker, Lieutenant ; and fifty-six Privates.

Captain Walter Pye's Company : Robert Parker, Lieutenant ; William Nathaniel French, Lieutenant. One Captain, three Lieutenants, four Sergeants, four Corporals, two Drummers, and seventy-six Privates.

Major Walter's Company : Joseph Harvey Bellass, Lieutenant ; James Gill, Ensign. Three Sergeants, three Corporals, two Drummers, fifty-six Privates, of whom apparently only twenty were present.

Captain Joseph Blagrave's Company: William Hall Timbrell, Lieutenant.

August 11th. There were nine companies at Lenham Camp. Another Muster Roll of the regiment is almost a duplicate of this one. Among other officers there was a Chaplain.

1782. Letter dated from Maidstone, written by Major John Walter, apologising to the Muster Master at the General Offices of the Horse Guards, that the Muster Rolls of the Berkshire Militia have been complained of as incomplete.

" To Officer commanding the companies of Berkshire Militia at Sevenoaks.

"It is His Majesty's pleasure that you cause the three companies of the Berkshire Militia under your command at Sevenoaks and Westonham to march on Saturday, the 18th inst., to Wrotham and Malling, where they are to be quartered till Monday, the 20th, when one of the companies is to march to the places mentioned in the margin hereof, where they are to be quartered and remain until further order. You will also acquaint this office with the receipt and execution of this order.—Given at the War Office this 14th day of May, 1782."

East and West Farleys, Barming, Teston, Loose, Boxley, Boughton and Linton.

" To Officer commanding the companies of the Berkshire Militia at Cranbrook, &c., Kent.

" It is His Majesty's pleasure that you cause the companies of the Berkshire Militia under your command at Cranbrook and Goudhurst to march on Tuesday, the 21st instant, to Maidstone, Aylesford and Berstead, where they are to be quartered and remain until further order. You will also acquaint this office with the receipt and execution of this order.—Given at the War Office this 14th day of May, 1782."

" To Officer commanding the company of the Berkshire Militia at Tunbridge.

" It is His Majesty's pleasure that you cause the company of the Berkshire Militia under your command at Tunbridge to march from thence on Wednesday, the 22nd instant, to Maidstone, where they are to be quartered and remain until further order. You will also acquaint this office with the receipt and execution of this order.—Given at the War Office this 14th day of May, 1782."

" To Officer commanding the companies of the Berkshire Militia at Tunbridge Wells.

"It is His Majesty's pleasure that you cause the three companies of the Berkshire Militia under your command at Tunbridge Wells and adjacents to march from thence on Friday, the 24th instant, to Maidstone, where they are to be quartered and remain until further order. You will also acquaint this office with the receipt and execution of this order.—Given at the War Office this 14th day of May, 1782."

" To Officer commanding the Berkshire Militia at Maidstone, &c.

" It is His Majesty's pleasure that you cause the Berkshire Militia under your command to march from their present quarters and encamp on Coxheath, on Monday, July 1st next, where they are to remain until further order. You will also acquaint this office with the receipt and execution of this order.—Given at the War Office this 24th day of June, 1782."

" To Lieut.-General Pitt, commanding the Berkshire Militia at Coxheath Camp.

" It is His Majesty's pleasure that you cause the Light Infantry Company of the Berkshire Militia under your command at Coxheath Camp to march on Tuesday, the 12th instant, to Rochester and Strood, where they are to be quartered and remain.—Given at the War Office this 7th day of November, 1782."

" To Lieut.-General Pitt, commanding the Berkshire Militia at Coxheath Camp.

" It is His Majesty's pleasure that you cause the eight companies of the Berkshire Militia under your command, which shall be remaining at Coxheath Camp on Friday, the 15th instant, to march from thence on that day, by the southern and Midland counties route, to Rochester, Strood, Finsbury, Chatham, Brompton and Gillingsham, where they are to be quartered and remain with their

Light Infantry Company already there until further order.—Given
at the War Office this 9th day of November, 1782:"

"To Officer commanding the Berkshire Militia at Rochester.

"It is His Majesty's pleasure that you cause the Berkshire
Militia under your command to march from their
present quarters in two divisions, according to the
routes annexed to the places mentioned in the
margin hereof, where they are to be quartered and
remain until further order.—Given at the War Office this 25th day
of November, 1782.  By His Majesty's command, in the absence
of the Secretary at War, M. LEWIS."

*Newbury,
Speen, and
Speenhamland.*

"Route for the five companies of the 1st Division of the Berk-
shire Militia: Thursday, November 28th, Dartford and Crayford;
Friday, November 29th, Greenwich, Deptford and Hatcham;
Saturday, November 30th, Hounslow and Cranford Bridge;
Sunday, December 1st, halt; Monday, December 2nd, Maiden-
head and Maidenhead Bridge; Tuesday, December 3rd, Reading;
Wednesday, December 4th, Newbury, Speen and Speenhamland,
and remain.  Route for the four companies of the 2nd Division
of the Berkshire Militia:  Friday, November 29th, Dartford and
Crayford; Saturday, November 30th, Greenwich, Deptford and
Hatcham; Sunday, December 1st, halt; Monday, December
2nd, Hounslow and Cranford Bridge; Tuesday, December 3rd,
Maidenhead and Maidenhead Bridge; Wednesday, December
4th, Reading; Thursday, December 5th, halt; Friday, December
6th, Newbury, Speen and Speenhamland, and remain."

December 3rd and 4th.  The Berkshire Militia marched
through Reading *en route* from Coxheath Camp to Newbury,
where they went into winter quarters.  On their march they
were met by His Majesty and three of the young Princes,
near Colnbrook, who did the regiment the honour of marching with them on foot from thence to Salt Hill.

After the Camp at Coxheath broke up, the following
letter of thanks was issued to the regiments which had
been quartered there:—"Lieut.-General Pitt cannot let the
Army separate without expressing to them in public orders

and in the strongest terms the great satisfaction he has
received, from the strict discipline and good behaviour
which the troops in general have maintained throughout
the whole of this campaign, and which redounds so much
to the credit of the corps that have been encamped here
this summer. The harmony that has so particularly prevailed through the whole Army during the campaign has
afforded the General the greatest pleasure, and he begs
the several corps to accept his best thanks for the great
attention and readiness which they have unremittedly
exerted upon all occasions.

1783. February. News was sent to the *Gentleman's
Magazine*, from Wantage, that Captain Price of
the Berkshire Militia, when returning home to Ham, was
set upon in the night of Sunday, January 19th, near
Wantage Churchyard, by two of the Yorkshire Volunteers
quartered in that town, one of whom had a hatchet, demanding his money and ordered him to turn out his pockets;
but the Captain springing from him, ran back to Ensign
Watson's lodgings, with whom he found Lieut. Banbury,
of the 66th or Berkshire Regiment, these offered their
services to go in search of the villains. They had not on
regimentals; but one gentleman had a sword, which he hid
under his coat. At the end of the town they were rushed
upon by the same men, one of whom again produced the
hatchet, which was seized, as well as the man, by Captain
Price and Ensign Watson, while the other was secured by
Lieutenant Banbury and both carried into safe custody.
Next day the commanding officer delivered them over to
the civil power, by whom they were committed to the
County Gaol. The names of the men were Robert and
William Brown. They were brothers and had two other
brothers in the same company, and were all four quartered
in the same house.

"To Officer commanding the Berkshire Militia at Newbury.

"It is His Majesty's pleasure that you cause the several companies of the Berkshire Militia under your command to march to such place or places within the county as you shall judge most convenient, to carry into execution His Majesty's orders for their being disembodied.—Given at the War Office this 4th day of March, 1783. By His Majesty's command, GEO. YONGE.'

Tuesday and Wednesday, March 11th and 12th. The Berkshire Militia were disembodied in the different towns in the county. By His Majesty's command, the men have all their clothes, knapsacks, &c., with their bounty arrears and fourteen days' pay, to carry them to their respective homes. Their excellent discipline and good conduct whilst on service, and the regularity of their behaviour to the last moment, entitle them to the highest praise.

1785. November. The War Office sent out orders to the Lord-Lieutenants of counties to put the Militia in order, and to send lists of officers, subalterns, and private men; also the state of arms, accoutrements and clothing, and the deficiencies of each corps. The Militia were, it was rumoured, to be embodied for one month; the first time this had taken place since the Peace of March, 1783.

1786. September 20th. A public breakfast was held at Sunninghill Wells, apparently to celebrate the King's escape from assassination. A Band was present from the camp at Sunninghill, but it is not mentioned to what regiment it belonged, probably from Windsor Garrison.

Twenty-sixth, George III., cap. 107, Act for concentrating Militia Laws, appoints Lords-Lieutenants of counties to raise and have the chief command of the Militia, and to appoint twenty Deputy-Lieutenants. This Act fixed 560 men to be raised in Berkshire.

1787. April. Thomas Hodgson, of Wantage, age 26, was convicted of robbery and executed at Ipswich.

He confessed to having enlisted in regiments in England, Scotland and Ireland, under forty-nine different names ; having often enlisted in different recruiting parties of the same regiment, he seldom stayed more than a day before deserting, yet he was only convicted three times of desertion and whipped once for it. He got 397 guineas in bounty money and fifty-seven guineas by robberies. I quote this story to show the difficulties experienced by regiments when enlisting recruits.

May 7th. An advertisement was issued, calling the Militia to come out for twenty-eight days' training at the Market Place, Newbury, in the name of the Lord-Lieutenant, Lord Craven. It and all succeeding advertisements are signed by James Payn, Clerk of the General Meetings, Maidenhead. The deserters were fined £20, or six months' imprisonment.

1788. May 12th. The Militia were ordered to assemble in the Market Place, Reading, at noon, for twenty-eight days.

1789. May 4th. To assemble in the Market Place, Newbury, for twenty-eight days.

1790. May 10th. Market Place, Newbury, for twenty-eight days.

1791. May 16th. Market Place, Newbury, for twenty-eight days.

1792. The regiment assembled in the Forbury, Reading, by order of the Lord-Lieutenant, Lord Radnor.

# CHAPTER IX.

## WARS AND RUMOURS OF WARS.—1792-1803.

TWO-THIRDS of the Militia were ordered to assemble in the Forbury, December 18th, 1792. A week later an order was issued to embody the whole Regiment, and deficiencies were to be made up by ballot, but the men came forward well and readily.

Clubs were formed, out of which substitutes could be drawn. This became a complete system of insurance against compulsory military service.

With regard to substitutes, the system was thoroughly over done, the sums paid sometimes, especially for foreign service, were absurd. In one case, the *Annual Register* says, £60 was given to a substitute, and for foreign service the prices asked were much higher. Parochial authorities preferred paying sums of money for substitutes to supporting the wives and families of soldiers absent during their term of service. Government allowed a certain amount towards the support of soldiers' wives and families, to endeavour to render the service less distasteful. In all cases, where it was possible, voluntary enlistment was encouraged by liberal bounty money and other means.

1793. January. Two meetings of the Lieutenancy of the county were held at the "Crown" Inn, Reading, to consider the Act of Parliament for embodying the Militia. Afterwards the districts were divided into sub-committees.

January 28th. Allowances were made to Militiamen's wives and families during embodiment by the magistrates. An indignant question on the subject had been asked in Parliament a fortnight previously.

February 4th. Commissions granted by the Lord-Lieutenant: James Wyld, Lieut., *vice* Henry Blackstone, promoted December 19th, 1792; Ensign Arthur Annesley Powell, Lieutenant, *vice* James Baker, resigned December 20th, 1792; Stanlake Batson, jun., Ensign, *vice* Francis Simpson, promoted December 20th, 1792; John Thomas Newbolt, Ensign, *vice* James Wyld, promoted December 28th, 1792; Charles Morice, *vice* Anthony Annesley Powell, promoted December 29th, 1792.

July 8th. The Berkshire Militia were encamped at Waterdown, near Tonbridge Wells. They were justly pronounced one of the highest appointed and best disciplined regiments in His Majesty's service. Previous to the men leaving the towns on the sea coast, where they had been quartered for five months, a general and noble entertainment was given them by the inhabitants as an acknowledgment of the universal good character they had obtained and the affection the inhabitants had formed for them. At Hastings 400 of them sat down at one table, where His Majesty's health and prosperity to Hastings, were drunk with repeated shouts of loyalty and affection.

February 11th. Yesterday the route for the Berkshire Militia was given. They were ordered to the coasts of Kent and Sussex. When the order was given the men gave three cheers. They were a fine regiment, in complete order for service. Their excellent conduct there ensured that they would serve their king and country well, wherever they were sent. They began their march on Tuesday, February 14th.

February 18th. Militias from inland were ordered to the sea coast to relieve such regulars as were destined for other service. Berkshire is not especially named, though it is said, the Oxfordshire Militia was eager for service. This warlike zeal of the Oxfordshire Militia had a disastrous ending some two years later, when they mutinied at Blatchingdon.

A meeting of the Lord Lieutenants of several counties in England and Wales was held at the " St. Albans " Tavern, in March, for the purpose of drawing lots to determine the precedency of the Militia of the said counties, during the continuance of the war.    The following numbers were drawn by the respective Lords-Lieutenant attending, or by the persons appointed for that purpose : Bedford 42, Berks 30, Bucks 38, Cambridge '11, Chester 16, Cornwall 34, Cumberland 20, Derby 26, Devon 41, Dorset 43, Durham 10, Essex 21, Gloucester 8, Nottingham 15, Oxford 9, Salop 28, Somerset 40, Southampton 6, Stafford 27, Suffolk 19, Surrey 18, Sussex 24, Warwick 31, Hereford 25, Hertford 44, Huntingdon 12, Kent 1, Lancaster 37, Leicester 2, Lincoln 3, Middlesex 22, Monmouth and Brecknock 14, Norfolk 4, Northampton 45, Northumberland 23, Westmoreland 29, Wilts 35, Worcester 36, Yorks West Riding 39, Yorks North Riding 33, Yorks East Riding 32, Carmarthen 17, Denbigh 7, Glamorgan 5, Montgomery 13.

March.    Commissions granted by the Lord Lieutenant of Berkshire : William Viscount Ashbrooke to be Captain, *vice* Henry Blackstone, resigned ; Henry Boyle Deane to be Ensign.

March 6th.    The Right Honourable Lord Barrymore conducting a number of French prisoners from Rye to Dover by the Berkshire Militia under his command, the whole party halted at the turnpike at the top of Folkstone hill ; after taking some refreshment, and on regaining his seat in his vehicle, a fusee which he carried with him went off, and shot him through the head.    He died in a few minutes, and so finished a short, foolish and dissipated life which had passed very discreditably to his rank as a peer, and, still more so, as a member of society.    So said the newspapers of the day ; but let us charitably hope poor Lord Barrymore was not so black as he was painted.    He came into his title and money very young ; and, no doubt, was surrounded by temptations, and led on by bad companions and bad advice. He was born August 14th, 1769.

April. Commissions: Stanlake Batson to be Lieutenant, *vice* Lord Ashbrook; John Thomas Newbolt to be Lieutenant, *vice* Earl Barrymore, deceased; Henry Pincke Lee, of White Waltham, to be Ensign, *vice* Arthur Annesley Powell, promoted; Henry Robert Ince, of Westminster, to be Ensign, *vice* Hon. Augustus Barry, resigned.

There was extraordinary heat all that spring and summer. Our soldiers were not then thought and cared for in the way they now are, so that violent extremes of heat or cold were serious matters.

August 12th. The army encamped at Waterdown, marched at nine o'clock and reached Ashdown Forest at two o'clock, where they pitched their tents. The ground had been prepared for them.

The Prince of Wales and his regiment, the 10th Dragoons, were camped at Shoreham at this time.

August 19th. Commissions in the Berks Militia, signed by Lord Lieutenant: Ensign Henry Boyle Deane to be Lieutenant, *vice* John Wallis, resigned; Ensign Henry Pincke Lee to be Lieutenant, *vice* Francis Simpson, resigned; Megi Henry Gilbert Stephens, gentleman, to be Ensign, *vice* John Edward Madocks, resigned.

In the *Reading Mercury* there is a rather curious advertisement for two deserters, Thomas Marsh and Richard Rider, giving full descriptions of them, and five guineas reward each for recovery. They belonged to Lord Craven's Company of Foot, commanded by Lord Paget. No doubt they had pocketed their bounty money and decamped, only to re-enlist elsewhere, a trick often played by rogues who thus gained large sums of money.

The Berkshire Militia deserters were never advertised for, but when caught were sentenced according to the law, administered by the Magistrates at the County Bench.

Throughout the French scare great excitement prevailed. Wooden Barracks were erected at Woolwich, Deptford, and other places, and camps were established all along the English coasts, especially in the South, the quarter where foreign invaders were most likely to land.

The Militia return for the Kingdom gave 36,602 effective men.

In December, four Companies of the Berkshire Militia marched from Brighton to Winchester, halting there one night. The following morning they proceeded to Romsey, a distance of 11½ miles, for the winter.

December. Commissions signed by Lord Lieutenant: Walter Pye, of the Temple, Esq., Major; Bernard Brocas, of Reading, Esq., Captain; Megi Henry Gilbert Stephens, Esq., Lieutenant; Charles Imhoff, of London, Ensign; Thomas William Ravenshaw, of Bracknell, Ensign; Swann Hill, of London, gentleman, Ensign; Charles Joseph Meter, of New Sarum, gentleman, Ensign.

October. The Camps were to break up October 20th. The Berkshire Militia were ordered to Southampton and Romsey. The Oxfordshire Militia were at Reading then, but a week later there was confusion, for instead of marching to fresh quarters as ordered, the order was cancelled, though the troops were all packed ready to start, for it was rumoured the French were collecting at Cherbourg for the invasion of England. The only thing known in camp apparently, was that forage for a fortnight was ordered, but all quarters were changed and every preparation was made for immediate defence in case of the French army landing on our shores.

The Reading Musical Society subscribed five guineas, three of which were sent to troops abroad under the Duke of York, and two guineas to the Berkshire Militia. This was to provide flannel waistcoats and warm clothing for the soldiers. During October, November and December of the

# THE BERKSHIRE MILITIA MARCH,

Written expressly for the Regiment by command of the

## COLONEL THE RIGHT HON. EARL OF RADNOR,

in 1792, by

Z. WYVILL, OF BRAY, NEAR MAIDENHEAD,

Harpsicord and Music Master.

*Copied from a collection of old Marches in the British Museum.*

above year, subscription lists were kept open at the various Berkshire County Banks, and a weekly advertisement of it inserted in the *Reading Mercury*.  It was a hard winter, and on the high downs, where the Camps were situated, the cold was intense and the men must have suffered very much.

1794    During February there was deep snow in England, both north and south, and the mail coaches were stopped and blocked with the drifts.    ·

In this year there was published a " March for the Berkshire Militia, composed for them by the desire of the Right Hon. Earl of Radnor, by his obedient servant, Z. Wyvill, of Bray, near Maidenhead."  It was written in score for eight parts : two clarionets, two oboes, two horns, and two basses. It was also arranged for the harpsichord.  It was printed by Longman and Broderip and sold by them, also by the Author at Bray, and by Smart and Cowslade, Reading, and was entered at Stationers' Hall.  The price was one shilling.

This primitive band was evidently kept up until the regiment was disbanded in 1816; for Alderman Darter, in his " Reminiscences," mentions the horns and also the " Serpent."  This last was a brass instrument like a horn. The band was evidently thought much of, and in 1798 numbered twenty performers.

The invasion of England by the French was again considered imminent, and in February the English were thoroughly frightened.  It was said 50,000 men were ready to cross the Channel ; Hastings being their chosen landing place.  Orders were at once issued to the Regulars and also to all Militia regiments, both British and Irish, to recall all men on furlough, except such as were on recruiting service, to rejoin their regiments immediately.  Every officer disobeying this order would be superseded, and non-commissioned officers and privates would be treated as deserters.

March 4th.  A meeting was held in the Grand Jury Room, Reading, to consult about the plan of " Augmenting the Forces for Internal Defence." A subscription was begun, and in the room £1,450 was subscribed; the names of the subscribers are given in the *Reading Mercury*.

The list was kept open some weeks, and a large sum of money collected, which went to pay for the Volunteers' Associations and Cavalry Regiments then raised.

Mr. Pitt brought forward a motion in Parliament to bring a bill ,before the House of Commons, for augmenting the Militia as in the late wars. He did not specify whether by ballot or volunteering but seemed to wish it left to discretion, rather than decided by a hard and fast law.

Early in March, a meeting was held [place not stated] of Colonels of Militia, who were determined to increase the Militia one-fifth beyond its present establishment by an additional number of privates to each company, but whether these men were to be raised by fresh ballot or by volunteer companies, as was done in the last war, was not immediately decided.

March 31st. A meeting was held by special appointment at the office of the Field-Marshall, the Duke of York, of two Militia Colonels and two Fencible Colonels, to decide the precedency of rank, which was determined in favour of the Fencibles.

The officers on half-pay, to whom the Secretary of War did not find it convenient to give posts in the Army, were invited by an official advertisement to accept similar stations in the Militia corps, without injury to their present rank. It is doubtful if this regulation met with success: no mention occurs in the *Gazette* of any officers availing themselves of it.

Among all the terrors of invasion, it is a pleasant change to meet with a little romance. It is seldom the Colonel of a Militia regiment marries during the embodiment of his regiment. But the following announcement gives: "Married, November 15th, 1794 (by special licence), at Norbiton Hall, Surrey, Edward Loveden Loveden, Esq.,

Colonel LOVEDEN.
1794.

M.P. for Abingdon, and Lieut.-Colonel of the Berks Militia, to Miss Lintall, only daughter of Thomas Lintall, Esq., late of Great Marlow, Bucks." Although it is not spelt as it is usual to see it, the name is recognisable as a well-known Berkshire one (that of Lenthall); but, in those days, spelling was still somewhat phonetic.

March 22nd. The *Gazette* contains the following order: "His Majesty appoints Colonels of the respective regiments of Militia to be Colonels in the Army by brevet, so long as their regiments shall remain embodied for actual service."

The appointment was made by the Right Hon. the Earl of Radnor, Lord Lieutenant of the County, of Mr. William Marsh, of Reading, to be agent to the Berks Militia in place of William Brummell, Esq., deceased.

Commissions signed by the Lord-Lieutenant: Edward Loveden Loveden, of Buscot Park, Lieut.-Colonel, *vice* Penyston Portlock Powney, Esq., deceased, dated February 20th, 1794; Ensign Imhoff, Lieutenant, *vice* Lieutenant Wyld, resigned January 24th, 1794; Pryse Loveden, of Buscot, Esq., to be Ensign, *vice* H. P. Lee, promoted; Thomas Stracy, of London, gent., to be Ensign, *vice* H. G. Stephens, promoted, dated February 25th, 1794.

At the end of April, the Government contracted with Mr. Augur, of Eastbourne, for ground for a small encampment at that place. It was to be formed of the South Devon and Berkshire Regiments of Militia, and the 11th Light Dragoons. The contract commenced from June 1st.

The Militia throughout the kingdom received orders from the War Office to be ready to take the field by May 10th.

Early in May the following curious advertisement appears in the *Reading Mercury*: "The public in general and the gentlemen of the County of Berks in particular are requested by the officers of the Berkshire Militia, which are at variance

K

with the Lord-Lieutenant of the County, to suspend their opinions relative to the difference at present subsisting between them until it shall be a proper time to lay a fair and clear statement of the whole transaction before them, as they fear it has been misrepresented."

The standard of height for the Supplementary Militia is given below. The question of bounty money given to Regular Soldiers, as contrasted with Militiamen, was a vexed question. It was said to draw men into the latter, to the detriment of the former.

"To all stout, able young men of 5ft. 4in. and upwards, without families, willing to enter the Militia for the county and to compose one company, which is to serve only with the Militia within the kingdom during the time they are embodied, or for a less time, if His Majesty should not think their services necessary, will, by applying to the Clerk of the Sub-division Meeting of their division, receive a bounty of £10, besides the marching guinea, and will enter into immediate pay and not be liable to march out of the United Kingdom."

April 26th. Yesterday sen'night, during divine service in the afternoon, a fire broke out in a bakehouse at Trotton, near Southampton, the owner of which had a contract for bread for the troops under Lord Moira. There was a prodigious pile of faggots near the house, but by the activity of the Berkshire Militia who were called out of church, the whole were removed before the flames had reached them; notwithstanding which, the bakehouse and a mansion adjacent were entirely destroyed. It is supposed the fire was occasioned by the oven being over-heated.

May 12th. All absent officers of the Militia in the United Kingdom were ordered to rejoin before the 14th.

May 19th. The Militia officers throughout England had His Majesty's permission to wear an undress of blue, with red cuffs and collar, on all service, except general field days and public reviews.

May 20th. On Tuesday, the Berks Regiment marched to Southampton; Wednesday, they proceeded to Wickham, where they halted; Friday, they moved to Havant and places adjacent; Saturday, they reached Chichester; Monday, they marched to Arundel; Tuesday, to Brighton; Wednesday, to Seaford; and on Thursday, to Eastbourne, where they were to be encamped.

June. Great excitement prevailed all over England over the British victory.

Joseph Blagrave, Captain of the Berkshire Militia, was appointed Captain Commissionary (? Commissary).

In this year, the war panic was at its height. Regiments were formed all over England. It seems as if anyone, who fancied, could apply for Royal permission to raise a corps. Every district rose in military ardour. In Berkshire alone, we have the "Windsor Foresters, the Loyal Berkshire Volunteers or the Reading Volunteers (this latter raised by Sir Charles Marsh, who was gazetted their Colonel soon after), the Wantage Volunteers (under Trevor Wheeler, Esq.), and the Abingdon Independent Cavalry;" and, from this period, commissions were formally notified in the various corps. By September, it was reported that a cordon of soldiers extended along the South Coast, of more than 30,000 men, chiefly Militia, Fencibles, and new corps; in November, the panic was subsiding, for they lowered the Army bounty money from twelve guineas to ten guineas. Everything took a military turn. Concerts were held in Reading, in which Military Bands played military pieces. The whole nation was inspired with a war fever, and the invasion, so long feared, seemed at last to have been imminent.

This was a year of tempests and torrents. A terrible gale swept along the coasts, beginning on Sunday, October 7th.

and lasting all through Monday.  At Dover it was the worst
gale since 1780; but, in those days, every storm was said
to be the "worst in the memory of man."  The camps,
of course, suffered much.  The account of the storm is thus
quaintly given: "The camps suffered greatly in the gale of
Sunday sen'night, by the violence of the wind which stript
a great number of officers and men of their marquees and
tents, and left many of them exposed to the rigour of the
elements, in a perfect state of nature; as soldiers, it should
be observed, do not always sleep in their shirts."

November.  The camp broke up and changed quarters
for the winter.  The Berkshire Militia, commanded by Lord
Radnor, were at Deal and Sandwich.

November 21st.  John Cohen, of the office of the Military
Society, 1, High Street, St. Mary-le-Bone, advertised that
malicious and untrue reports had been circulated, damaging
to the society, which he denied; but, his denial reads rather
as if it was put in the newspaper more as an advertisement,
to publish the fact that the society provided Militia substitutes, than for any other cause.

Following the gale of November, came incessant rain and
high floods.  At Loddon Bridge the water was so high, that
the coachman of the Forest Coach dared not cross, but
returned.  The flood on the Kennet was higher than known
for many years.

1795.  January.  Commissions signed by the Lord-
Lieutenant: Capt.-Lieutenant A. N. Toll to be
Captain, *vice* Joseph Blagrave, resigned; Ensign Stracy,
Lieutenant, *vice* A. A. Powell, resigned; Ensign Bardesley,
Lieutenant, *vice* H. P. Lee, resigned; Edward Reeves, of
Arborfield, Esq., Lieutenant, *vice* S. Hill, resigned; Lieut.
Hawes, Captain-Lieutenant, *vice* A. N. Toll, promoted;
Richard Weekes, of Barkham Square, Esq., Captain, *vice*
W. Viscount Ashbrook, resigned; John Newbury, of Heath-
field Park, Sussex, Lieut., *vice* Francis Hawes, promoted.

The Hawes owned Purley Hall, which they bought from the Hydes. They changed the name from Hyde Hall to Purley Hall. Hawes was a rich shareholder in the South Sea Company. (Not the notorious Hawes, though probably related). After the collapse of that celebrated company, the Hawes began to mortgage the estate, and the son of the last owner was, I believe, a linen draper in Reading, after the Wilders had bought the property in 1779.

March. There arose a question in the House of Commons as to the pay and clothing of the Militia; also as to the pay of Militia Subaltern Officers. General Tarleton, who replied, was evidently a man of advanced views, for he said he should never rest satisfied till the Militia was assimilated with the standing Army; and he alludes to the organisation of the German Army.

The war panic and fear of invasion was again active. This time the Dutch seem to have caused additional un-easiness, for the cordon of soldiers along the coasts had been doubled and amounted to 150,000 men.

The Duke of Richmond's district of forces was a large and scattered one. The division at Hythe Camp consisted of the Berkshire Militia, together with the Militias of Lancashire and North Devon, the new Romney Fencibles, and the Warwickshire Fencibles.

I shall give, if possible, the regiments actually encamped with the Berkshire Militia; because curious feuds and jealousies always, I believe, exist as traditions in all regi-ments, some probably dating back to this old war time; for it is said that certain regiments cannot be quartered together without attacking each other.

April. At Lewes Camp, a serious mutiny arose in the Oxfordshire Militia at Blatchington Barracks, near Seaford. Afterwards a Court-martial was held, composed of twelve officers, each drawn from a different regiment. The Berk-shire Militia was represented by Major Pye.

June 13th.  The following letter was sent to the printer of the *Reading Mercury* for insertion, by the High Sheriff of Sussex, from Brighton :

"Horsham, June 12th.

"I am now proceeding to the execution of the two poor fellows"

"June 13th, 1799.  At a quarter past nine.

"I am just returned from the ground where two soldiers of the Oxfordshire Militia were shot this morning, about a quarter past eight.  One of them knelt down upon one coffin and one upon the other, and they both instantly fell dead.  ·Though left there, lest there might be any remains of life, a firelock was let off close to the head of each immediately after.  The scene was the most awful and impressive I have ever seen.  It was in a valley about a mile distant from the camp whither all the troops, Cavalry, Infantry and Artillery, were drawn up in two lines; and, after three men out of the six who had been sentenced to be flogged had received their punishment in a very exemplary manner, the three others were pardoned.  The men capitally convicted were then marched up between the two lines of the Army, accompanied by a clergyman, and escorted by pickets from the different regiments of horse and foot; and at the upper end of the line, after a short time spent with the clergyman, they were shot by a party of the Oxfordshire Militia, who had been very active in the late riots, but had been pardoned.  The men appeared very composed and resigned; and the party who had shot them were, many of them, very much affected after.  Indeed, several men of the regiment seemed greatly agitated and concerned.  An example so unusual and so terrible will, it is hoped, have the desired effect upon the minds of the Militia, and shew the danger of using the arms which are entrusted to them, for the intimidation instead of the defence of their county.  The awful ceremony was concluded by the marching of all the regiments round the bodies of the unhappy soldiers as they laid on the ground."

The following week's paper gives a long account of the second executions, which were less formal.  The other rioters were marched off.  They preferred to submit to any punishment, but refused the compromise offered, which was to join the Royal American Regiment. Another man of the Oxfordshire Militia was executed shortly after, for robbery from a shop.

Mutiny seems to have been rife. The Army had been long in camp, and, in spite of severe punishments, rioting seems to have been general. The punishments given for offences were severe in the extreme. In some cases of disobedience and absence without leave, the sentence was 1,000 lashes; while 400 or 500 lashes were ordinary sentences for misdemeanour. Seventeen years before this time, a new regulation was passed, that all deserters from any of the Military corps were to be sent to the East Indies, or the coast of Africa, for life. But this regulation, which was read at the head of every regiment in Great Britain and Ireland by the King's order, was evidently unpopular and seldom, if ever, enforced.

June 21st. The night was so terribly cold, that hundreds of newly-shorn sheep were found dead next morning and a gale raged all along the Dorsetshire coast. These gales and floods caused a great scarcity; famine and riots were frequent. Corn was very scarce. The baker to the camp at Brighton made his best bread with one-third of potatoes, and found thereby that he saved four bushels of flour daily, while the bread gave equal satisfaction to the officers and men. The Prince of Wales ordered only brown bread to be served at his table; and the officers in camp at Brighton ordered the same, and enforced it under forfeiture of a month's pay from anyone who disobeyed the order.

August. Following the gales, was a heavy storm of thunder and lightning; but there was great rejoicing over the promise of a plentiful harvest, and agricultural matters seemed more to the fore than military ones. But this tempestuous weather must have been felt in the camps along the coasts.

A private in the Berkshire Militia encamped at Sandgate was, at this time, discovered to be a female. She
had served in the regiment for six years without discovery and with great credit, and was remarkable for her
cleanliness. Her father was boatswain of a man-of-war.
When his daughter was young, he put on her the breeches,
called her William, and entered her on board his own ship
as cabin boy. Some differences arising between them she
ran away, went into the farming line as keeper of sheep,
&c., and at last entered for a soldier. It was observed that
she always slept in her breeches and jacket, which were
close buttoned all round; but no suspicion was ever entertained of her sex, until a quarrel happened between her
and some comrades, and they threatened to cob her.
Through fear of the punishment, she disclosed herself to
the Sergeant-Major's wife. Her behaviour was so prudent
in the regiment, that the officers subscribed something
handsome to clothe her properly and carry her home. She
was about twenty-eight and of a comely appearance.

Duels had not ceased in 1799. Some were fought on the
most trifling disagreements. We read of two officers of the
North Lancashire Militia differing over how to cut up a leg
of mutton, and retiring to settle the dispute by six shots at
twelve paces; but, after all, only one was wounded in the
foot! The rule, apparently, was that the sender of the
challenge was the only one to fire, and his opponent either
did not fire at all, or fired in the air.

Autumn. Commissions in the Berkshire Militia: William
Lloyd, of Shrewsbury, Esq., to be Lieutenant, *vice* Richard
Weekes, promoted; Ellis Mears, of Southampton, gent., to
be Ensign, *vice* J. Maton, resigned; Henry Boyle Deane, of
Reading, Esq., to be Captain, *vice* W. Timbrell, resigned.

May. The Berkshire Militia marched from Dover to be
encamped at Shorncliffe, near Sandgate. The Warwickshire and East Essex Militias were encamped at the same
place.

October. Glanders broke out in Brighton Camp, and about fifty horses of the Prince of Wales' Regiment had to be shot.

October 11th. There was a grand review, held by the Duke of York, after which the camp broke up for the winter.

New regulations were issued to the Army. Every soldier in future was to receive 8d. per day, and the allowance of bread was to be discontinued.

October 28th. A most elegant ball and supper was given at Buscot Park, to celebrate Captain Pryce Loveden's twenty-first birthday.

Wednesday, October 21st. Three hundred of the Berkshire Militia marched into winter quarters at Margate. Ramsgate was the headquarters of the regiment.

Commissions: John Pocock, of Blewbury, Esq., Ensign, *vice* Pryce Loveden, promoted; John Hill, of Barkham, Berkshire, gent., to be Ensign, *vice* Swan Hill, resigned; David Crowe, of Sindlesham, Esq., to be Ensign, *vice* Thomas Stracey, promoted; Benjamin Holloway, jun., of Charlbury, Oxfordshire, to be Ensign, *vice* J. Bardsley, promoted. William White, who had been Regulating Captain for Berkshire, was promoted to Liverpool and given a good appointment.

October. The chief subject of interest was agriculture, the scarcity of corn and the price of bread, varied by debates on a dog tax, suggested through the prevalency of hydrophobia. For a time, warfare and military affairs were put aside for other topics.

Christmas. The reduction of the forces was steadily going on, soldiers were drafted into other regiments by degrees.

1796. January. Among the Military Cantonments, the Berkshire Regiment was in quarters at Ramsgate and in the Isle of Thanet.

February. Commissions signed by the Lord-Lieutenant of the County: Ensign Mears, Lieutenant, *vice* Stracey, resigned; Ensign Pocock, Lieutenant, *vice* Deane, promoted; Richard Fiennes Wykham, Esq., Ensign, *vice* J. Hill, resigned; Ensign Holloway, Lieutenant, *vice* Crowe, resigned; Ensign Wykham, Lieutenant, *vice* Mears, resigned; Charles Imhoff, Esq., Captain, *vice* Sheppard, resigned.

March. An alteration was proposed in the infantry dress, by which long coats were to be entirely excluded.

May 7th. Parliamentary duties prevented the Earl of Radnor from attending his regiment on their route from the Isle of Thanet into Devonshire. His Lordship, however, transmitted orders to his household at Longford Castle to entertain his men in the best manner circumstances would permit, when they passed that place; in consequence of which the first division, previous to their entrance into Salisbury on Tuesday sen'night, crossed the river in boats from Alderbury to Longford, and at the Castle were plentifully regaled with bread and cheese and good English beer. The second division, having more time to spare, marched on Friday at noon from Salisbury, and had the same fare spread on tables in the park, returning in the evening highly exhilarated. The third division was entertained in the same manner. To the credit of the corps, their hilarity produced not a single instance of disorderly conduct, but manifested itself in songs of loyalty and loud huzzas in honour of their noble commander.

May 21st. The Summer Cantonment of the Berks Militia was Totness and the adjoining towns.

December 28th. A meeting was held at the "Crown" Inn, Reading, to consider the augmentation of the Militia according to Act of Parliament. Present: Sir Francis Sykes, Bart., John Bagnall, Esq., Edward Golding, Esq., Henry Deane, Esq.

Monday, January 2nd. Wokingham division
1797. had to provide for two officers, two sergeants,
one drummer, nine rank and file of the old Berkshire
Militia, and 126 privates of the new Supplementary Militia,
for twenty days; and the sum of £126 was voted, to provide
for the subsistence of the 126 privates during that time.

A gradual reduction of the Army was made. The
Supplementary Militias were called out in detachments at
the different towns, for the space of twenty days; each
section was disembodied after training. Later on in the
year, when war seemed more imminent, these were allowed
either to join certain regiments sent on march to recruit,
or else to be drafted into the old County Militia.
Everywhere the Supplementary Militia gained the same
praise as the old Militia for good conduct; at Wantage,
especially, where they were entertained before breaking up.
Defaulters from the Supplementary Militia were advertised,
and non-appearance with the last section was punished as
desertion.

At Wallingford, Lieutenant Bardsley and the detachment
of non-commissioned officers of the old Berkshire Militia
who drilled the Supplementary Militia there, were spoken
of in the following terms, besides receiving a small present:
" For their good conduct and behaviour, as well in the town
as in the field, too much cannot be said in their praise; suf-
fice it to say they behaved themselves as soldiers and men."

February. The Berkshire Regiment was quartered as
follows: The Grenadiers, under Captain Weeks; head-
quarters, Dartmouth. Four companies in Barracks at Bury
Head, under Lord Radnor, Captain Brocas, Captain Imhoff,
and Captain Ravenshaw; the volunteer company at Brixham; three companies at Totness and Bridgetown, with
the Lieut.-Colonel, the Majors, and Captain Deane; the
Light Infantry, with Captain Toll, at Cawsand and adjacent places. This gives nine companies to the regiment.

It was at this time the extraordinary event happened of the French landing at Fishguard, on the coast of Pembroke-shire. The English were, however, too well prepared for them, and they were promptly made prisoners. It was reported that three of them were killed and 1,400 prisoners taken. They had three 52-gun ships, with 460 men; one ship of twenty-two guns, and a lugger of fourteen guns. But reliance on these statistics is doubtful, as later on I find the number of French prisoners in England variously esti-mated as 22,000 and 26,000; and, still later, it was said there were 4,000 Dutch prisoners also. What ever the exact number was, it evidently was a very large one.

Influenza was prevalent. The newspaper speaks of it as " that old disorder, known by the old term influenza." It took the form of a violent cold in the head, general lassitude and weariness through the whole frame, which was found extremely difficult to remove. No particular mention is made of this disease among the soldiers, doubtless they suffered from it at this time.

May. Lord Radnor was not with his regiment early in this month, as he was present at the General Quarter Sessions for Wilts.

During this month the War Office issued orders for strict enquiry to be made of the names of all Militiamen who had enlisted in Regular regiments and the regiment wherein they had been enlisted, so that they might be ordered to be given up to their own Militia. This was to avoid desertion and re-enlistment for the sake of bounty money.

July. Commissions in the Berkshire Militia: Edwin Reeves, of Arborfield, Esq., to be Captain, *vice* Gill, resigned; Francis Robert Holdsworth, of Dartmouth, gent., to be Lieutenant, *vice* Ravenshaw, promoted; Ensign Cane, to be Lieutenant, *vice* Imhoff, promoted; Robert Cane, of London, gent., to be Ensign, *vice* Pococke, promoted; Thomas William Butler, of Wokingham, to be Ensign, *vice* Crowe, promoted.

This year there was another terrible thunderstorm, felt especially at Plymouth and Weymouth. The Berkshire Militia were on the Devonshire coast, around Totness and Dartmouth.

September. The Government announced their intention of sending each Militia, for winter quarters, back to their own counties; but it was not carried out.

October 1st. The Berkshire Militia went into winter quarters at Bristol.

November. The civil department of every regiment of the Line, as well as the Militia, heretofore conducted by military men called Captain and Paymaster, was about to undergo a complete change. In future, persons conversant in agency affairs were to be appointed Paymasters of regiments, with the rank of Captain in the Army and 15s. per day as pay, without deduction. They were to be attached to the regiment, appointed by the Colonels, and approved or recommended by the Army Agents. Some security was to be taken in consequence of the appointment. The regulation came into force on December 15th. The plan was attributed to H.R.H. Duke of York, who, as head of the Army, interested himself in all details of its management, both great and small.

An erroneous account of the duel at Bristol between Lieut.-Colonel Sykes and Mr. C. F. Williams having appeared in the Bristol papers, the following was sent to the printer of the *Reading Mercury:*

"Immediately after Mr. Williams's apology, we, the undersigned seconds in the affair, think it proper to insert the following statement of facts: On the 11th inst., a letter was addressed to the printer of *Farley's Bristol Journal*, signed 'Trim,' which Colonel Sykes conceived reflected on his conduct. On enquiry of the printer, he was informed that Mr. Williams was the author; and the next day after the information, Colonel Sykes, meeting Mr. Williams in College Green, asked him if he was the author of that letter, and at the same time shewing him the paper. Mr. Williams replied in the affirmative, and the Colonel then asked him if he alluded to him in that letter. He replied he did in a part of it, when the Colonel immediately struck him several times. On the Friday morning following, a meeting took place between the parties, in consequence of a message received by the Lieut.-Colonel from Mr. Williams; and, at their arrival on the ground, it was agreed that Colonel Sykes and Mr. Williams should stand at ten paces distant and fire together on a word being given. The ground being measured, they took their posts and fired accordingly, when Colonel Sykes's ball passed through Mr. Williams's cravat, waistcoat, and the cape of his coat. Mr. Williams missed the Colonel. On the second discharge the Colonel received a ball through his foot, the Colonel's ball having passed close under the brim of Mr. Williams's hat. At the third fire Mr. Williams's pistol snapped, and the Colonel slightly wounded him in the groin. At the next discharge the Colonel was shot through the pocket of his coat and Mr. Williams was missed. The seconds then interfered, when it was settled with the consent of Mr. Williams that he should make Colonel Sykes an apology in the public newspaper; and, in consequence of that apology, Colonel Sykes should apologise to him in College Green, before two or three friends, for having struck him, as soon as his surgeon will permit him to come out. We think it necessary to add that, although the Colonel was shot through the foot at the second fire, he concealed it from us, and did not make it known until the affair was satisfactorily settled.—JOHN ALGOE, ROBERT CANE, seconds."

Then follows a letter of apology from Mr. Williams to Colonel Sykes, duly signed by the seconds.

Coloured feathers were worn by the officers and soldiers, to distinguish the companies; red and white belonged to the body of the regiment. The Grenadiers had white feathers and the light company green. The use of badges for the purpose did not come in until the early part of this century. Probably they superseded the feathers on account of the expense; for, with wind and weather, the latter soon became faded and shabby, and required to be often renewed.

1798. January. A plan of defence was drawn up by Government to protect the South coast. Fresh orders were issued. Among other regiments one was formed of boys only ; neither the name nor locality of this novel corps is stated. The papers were full of warlike rumours, and the King ordered out the Supplementary Militia, but only to allay public fear. The papers then said that this was not done in consequence of any information, but merely as a precaution, so that the men might be ready if required. The effective strength of the Military of Great Britain was reported to exceed 200,000 men.

February. The Militias were ordered back to their own county towns, and all officers on leave were ordered to rejoin. Many of the Supplementary Militia were drafted into Line regiments at this time.

February 19th. The officers, non-commissioned officers and privates of the Berkshire Regiment of Militia, in garrison at Bristol, have subscribed two days' pay to the voluntary contribution for the support of their country. The following was given in general orders on the occasion :

" Bristol,
"February 19th, 1798.

"Lieut.-General Rooke acquaints Colonel the Earl of Radnor and the officers, non-commissioned officers and privates of the Berkshire Regiment of Militia, that he shall have the greatest satisfaction in making known (through the Adjutant-General) to Field-Marshall H.R.H Duke of York the voluntary subscription of two days' pay each, for the support of Government."

The Supplementary Militia, its divisions and sub-divisions were again out on march. ·Half of them, 374 men, were balloted for. They were embodied at Newbury, and from there drafted to the main body of the regiment at Bristol.

February. Pryse Loveden, Esq., of Woodstock, was married to Hon. Mrs. Agar, sister to Lord Viscount Ashbrooke.

The Supplementary Militia were to be disposed of as follows : One-third were allowed to enlist, on considerable bounties, in several parties of marching regiments sent to recruit among them ; the rest were to join the old regiment of Militia, belonging to their respective counties.

Circular letters from the Duke of York were sent to Colonels of several Militia Regiments, directing them that three sergeants, three corporals and six privates (who were eligible for the rank of non-commissioned officers) should be sent from each regiment, to aid in disciplining the Supplementary Militia, from which corps they were to receive an equal number of men in return.

March. It was said 30,000 men were called out of Supplementary Militias.

March 14th. Two sergeants and eighteen men were sent to Bristol from Newbury.

April. One Field Officer, two Captains, four Subalterns, with non-commissioned officers and a drummer, were ordered from Bristol by May 3rd, to proceed to Newbury to assist in training the Supplementary Militia. The officers to travel in carriages, the non-commissioned officers by outside coaches, &c. Four days after assembly the men had the option of enlisting in the Army, receiving a bounty of seven guineas, to serve during the war.

Wednesday, May 23rd and 24th. The detachment of the Berkshire Supplementary Militia were ordered to join the regiment at Bristol. They were to march in two divisions, *via* Hungerford, Marlborough, Chippenham and Marshfield, halting on Sunday, the 27th, arriving at Bristol on May 28th and 29th. The following day another order was issued, directing them to start two days earlier, *i.e.*, 21st and 22nd, so as to arrive at Bristol at the end of the week.

June 9th. The Berkshire Militia were relieved at Bristol Garrison by the Royal Cheshire Militia, under Lord Grey. They marched through Bath to Poole, under Lord Radnor.

Commissions in the Berkshire Militia: Henry Bromley, of Caversham, Esq., Ensign, *vice* Gill, resigned; Benjamin Bailey, of Caversham, gent., Ensign; Ensign Ramsey, Lieutenant, *vice* Reeves, promoted; Arthur Vansittart, jun., of Shottesbrooke, gent., Ensign, *vice* Ramsey, promoted; John Blagrave Pococke, of East Hagbourne, Captain; Augustus Henry East, of Hurley, Captain; Ensign Dodd, Lieutenant, *vice* Pococke, promoted; Ensign Birnie, Lieutenant, *vice* Holloway, resigned; Ensign Guyenett, Lieutenant; Ensign Hill, Lieutenant.

June 11th and 12th. The Berkshire Militia left Bristol. The regiment consisted of twelve companies, and marched by two routes. The first half divided into two portions of three companies each. Three companies started on Monday, the 11th, for Bath; Tuesday, the 12th, Warminster; Wednesday, the 13th, Shaftesbury; Thursday, the 14th, they halted; Friday, the 15th, Blandford; Saturday, the 16th, Poole Barracks. The three companies of the second division left on the 12th, and went exactly the same way, except that they reached Blandford on Saturday and spent Sunday there, getting to Poole Barracks on Monday, the 18th. These six companies remained at Poole until August 27th, when they started for Weymouth, *via* Wareham, and arrived the following day. The other half of the regiment also left Bristol on June 11th and 12th, for Shepton Mallet, Wincanton and Sherborne, halting on Thursday, the 14th, in the two last-named towns, and then to Dorchester, and Weymouth town and Barracks on June 16th and 18th.

The Bill for allowing the Militia to serve in Ireland was passed; and our Militia evidently volunteered at once for service, for the *Reading Mercury* announces that the Berk-

L

shire Light Dragoons landed in good health at Pigeon House, Dublin. Curiously there is no mention in the marching orders of their going to Ireland. Alderman Darter refers to it. He tells of one of the Militia Band, the player of the "serpent," who lost his eye in Ireland and had to leave the regiment; and he also speaks of a boy who served in the regiment in Ireland, but I can find no other entries and they are not named among the first list of regiments sent to Ireland. Evidently the horse troop only was sent for a few weeks, while the body of the regiment remained at Bristol. We know there was a mounted troop belonging to the Berkshire Militia by the following advertisement, inserted in the *Reading Mercury* of July 23rd, ordering the embodiment of the Provisional Cavalry: "To assemble on such mares or geldings as had belonged to it, or substitutes." The allowance per man was three shirts, two pairs of stockings, and one pair of shoes. They were to assemble at Newbury on August 2nd, at twelve noon. This order was signed by the Earl of Radnor and dated from Poole, July 10th, where the Militia was then stationed. The substitutes for the Reading Provisional Cavalry were required to meet on August 2nd, on Theale Common, to join the remainder of the troop. This is signed by J. Blagrave, Major. As Major Blagrave was in the Berkshire Militia, it looks as if he was "lent" to the Provisional Cavalry; unless this was Joseph Blagrave, who had resigned in 1795. This cavalry appears to have formed part of the Militia and not part of any association for volunteers. They proceeded to Bristol to join the regiment, under the command of Major Stead ; and it is specially remarked that they were as steady, orderly and well-behaved as old soldiers, and not a complaint was made since their embodiment. This troop of horse must have been raised on the same lines as the troop of Militia Horse of 1660. They did not arrive at Bristol until August 20th. So they were not the same mentioned as sailing to

Ireland in June; but, perhaps, did duty while the Dragoon company were in Ireland. No other Militia regiment, so far as I have studied their histories, had a mounted troop, but the Berkshire Militia undoubtedly had.

This is the last entry I can find relating to a mounted troop in connection with the Berkshire Militia.

July. An order had gone from the War Office to the General officers commanding in the Eastern Division, to allow one-third of the old privates of Militia regiments to go on working-furlough during the harvest. This order probably was general and explains the number of absentees from each regiment.

August 16th. A general meeting of the Lord-Lieutenant and Justices of the Peace was held at the "Pelican" Inn, Speenhamland, to consider the Militia Acts. They dealt principally with the Supplementary Militia and condemned the Constables as ill-performing their duties. It was re-marked that the Reading Division and the Borough of Reading were grossly defective.

The number of duels between officers in the Army seem to have increased about this time. None in the Berkshire Militia are recorded, so we may presume they were a peaceable lot.

September. More Militias volunteered for service in Ireland and were forthwith sent.

Their Majesties, George III. and Queen Charlotte, with their daughters, arrived at Gloucester Lodge, Weymouth, after five o'clock on the evening of Saturday. On entering the town they were received by a party of the 3rd Light Dragoons, the Berkshire Militia, and flank companies of the Shropshire, North Hants, North Gloucestershire and South Devon Militias, the Weymouth Volunteers, and the Wyke Independent Fusiliers.

September 18th. The Earl of Radnor's Regiment of Berkshire Militia was drawn up and reviewed by the King,

L 2

in the presence of a large number of spectators. The troops went through all their manœuvres, which lasted upwards of three hours, and with great credit to their noble commander. The corps consisted of 1,000 men and made a fine appearance. After the review, His Majesty rode to the Infantry Camp on the Nore. The Royal Family partook of an elegant repast at Lord Radnor's house on the Esplanade, on their return to the Lodge, prepared for this occasion by the Countess of Radnor, who waited on their Majesties and the Princesses. Lady Radnor was a personal friend of the daughters of George III. and attended them in some of their tours.

Saturday, September 24th. At 10 o'clock in the morning the Berkshire Militia, under the command of Lord Radnor, was again reviewed by the King, after which the Royal Family partook of a public breakfast, given by his Lordship, near the old Castle. His Majesty was pleased to bestow the highest commendations on the soldier-like appearance of the regiment, and was himself in highest spirits. In the evening the Royal Family went to the Theatre, which was filled in every part at an early hour.

The news of Nelson's victory at the Battle of the Nile, fought on August 1st, caused great excitement everywhere. The King caused the Admiral's letter to be read publicly. The troops from the camp at the Nore fired a *feu de joie* in honour of the occasion. These were the 3rd Dragoons, flank companies of the Royal Cheshire and Berkshire Militias, and the Weymouth Volunteers. After firing, the troops marched past the King and the Duke of York. In the evening the Royal Family attended the Theatre. An address was given on the victory, and the Band of the Berkshire Militia, numbering nearly twenty performers, marched to the front of the stage and performed the chorus of "God save the King;" after this, they played "Rule Britannia," amid the plaudits of the audience. The King and Royal

Family were decorated with branches of oak and laurel, and the Royal Princesses had bouquets of the same.

October 27th. Four companies of the Berkshire Militia marched from Weymouth for Portsmouth, and four more companies had taken the same route the previous day; the remainder marched this day.

December. Commissions in the regiment of Berkshire Militia: William Lawrence Brookman, Ensign, *vice* Dod, promoted, dated June 14th; Hon. Henry Bromley, Lieutenant, June 29th; Arthur Vansittart, jun., Captain, July 6th; Ensign Benjamin Bailey, Lieutenant; George Henry Vansittart, Major, July 15th; Francis Robert Holdsworth, Captain, October 18th; Ensign George Treacher, Lieutenant, *vice* Cane, resigned.

On December 20th was passed the Act enabling His Majesty to accept the services of any of the Militia who voluntarily offered to be employed in Ireland.

As Militia regiments had been sent to Ireland in that Autumn, the Act passed in December, 1798, must have been the formal confirmation by Parliament of permission to serve out of England.

1799. The Militia expenses, as granted by Parliament this year, amounted to over four million pounds.

Commissions signed by the Lord-Lieutenant: George Henry Vansittart, to be Lieut.-Colonel, *vice* Sykes, resigned August 24th; Hon. Henry Bromley, to be Captain, *vice* Imhoff, resigned February 4th; Edward Martin Atkins Atkins, to be Captain, *vice* Deane, resigned June 5th; — Brookman, to be Lieutenant, *vice* Houldsworth, promoted December 13th, 1798; Ensign Macpherson, to be Lieutenant, *vice* Dodd, resigned February 4th; Ensign Schrader, to be Lieutenant, *vice* Hill, resigned February 5th; Ensign Guy, to be Lieutenant, *vice* H. Bromley, promoted March 11th; Ensign Sherren, to be Lieutenant, *vice* Guyenett, resigned May 18th; Lachlan Macpherson, to be Ensign, *vice*

Birnie, promoted December 17th, 1798; Thomas Key, to be Ensign, *vice* Guyenett, promoted December 18th, 1798; Frederick Henry Schrader, to be Ensign, *vice* Hill, promoted December 19th, 1798; George Guy, to be Ensign, *vice* Bromley, promoted January 19th; William Sherren, to be Ensign, *vice* Vansittart, promoted February 4th; Thomas Hughes Edwards, to be Ensign, *vice* Bailey, promoted February 13th; Thomas Snook, to be Ensign, *vice* Treacher, promoted February 15th; Robert Yeates, to be Ensign, *vice* Brookman, promoted February 21st; William Collis, to be Ensign, *vice* Macpherson, promoted February 25th; Charles Houlden Walker, to be Ensign, *vice* Schrader, promoted March 11th; Benjamin Smith, to be Ensign, *vice* Guy, promoted June 5th.

January. The Secretary for War determined to do away with the post of Regimental Chaplains, and, as a beginning, reduced their salaries. As in all matters, while embodied the Militia regiments were the same as Line regiments, this regulation also affected them.

June. The flank companies of the Berkshire Militia, which were in Winchester Barracks, marched to join the body of the regiment in Portsmouth Barracks. As usual, the Berkshire Militia met with the warmest praise. When it became necessary to strengthen the regiments of the Line, it was offered to the Militia regiments to volunteer into them. At this crisis, 333 privates and six officers of the Berkshire Militia volunteered at once, without the slightest solicitation; but only 263 privates and four officers could be accepted, as the Act of Parliament limited the number which could be drawn out of one regiment, only three-fifths being allowed to volunteer out of each regiment, whereupon several non-commissioned officers volunteered to serve as privates. Later on in the year, October, the orders permitting the transfer of soldiers from Militia to Line regiments to join their brave comrades in Holland, were read on parade; and 150 of the Berkshire Militia immediately turned out to serve in any part of the world with Captain Holdsworth, and they were attached to the 15th Foot. This transfer of men to the Line regiments reduced the old Militias so much, that recourse to the ballot became necessary.

The Supplementary Militias all over England were disembodied: 150 of the Berkshire Militia were thus dispersed to their homes from Reading.

1800. July. The Berkshire Militia then in Netley Camp were less strong by 500, owing to this cause, than the last time they were at Southampton.

From the Adjutant-General's return of the Forces, the effective Military were 191,452, exclusive of the Marines, who were in the Admiralty Department, and the numerous Volunteer corps who did not receive pay.

September. A detachment of the Berkshire Militia were in Winchester, escorting French prisoners. A riot was expected at New Alresford, it being market day, and much agitation everywhere on account of the price of corn and consequent rise in the price of bread. The Fawley Light Dragoons were summoned from Winchester, and the Berkshire Militia volunteered through their officers to help and were accepted; likewise a detachment of the Cornish miners. As evening approached, the crowd thickened, farmers and millers were hissed and the disorder increased. The magistrates tried to keep the peace but without avail, and were obliged to read the Riot Act. At last the Dragoons were ordered to charge; this they did with alacrity, and the crowd disappeared. In a few moments the street was empty. Many of the rioters were severely beaten about the head and back. Nine of them were taken into custody. No one was mortally wounded, for the Dragoons used only the flat of their swords; but they vowed that the next time such a scene occurred, they would fire first and charge in real earnest.

December. There died at Hannay, near Wantage, a private of the Berkshire Militia, named William Spindle, who was home on furlough. He was given a military funeral by the Wantage Volunteers, who, with great credit to themselves, gave him those honours due to every brave soldier. They marched with their band. The procession was most solemn, and drew tears from the eyes of the vast number of spectators who assembled to see it.

Unfounded rumours appeared in general newspapers, that Government intended to disembody the Militia; but this was far from being the case, as the War was not over, and especially at this crisis, the Militia was needed for home defence, many regiments being yet absent on foreign service.

1801. April. At the end of the month the Berkshire Militia marched into Lymington, from Portsmouth Barracks, when the commanding officer received the following extract from General Orders at Portsmouth:

"Headquarters, Portsmouth, April 21st, 1801.

"Sir,—I am directed by Major-General Whitelocke to forward to you an extract of the General Orders of to-day:

"To Lieut.-Colonel Ravenshaw, commandant of the Berkshire Regiment.—General Orders: Major-General Whitelocke, having made the half-yearly inspection of the Berkshire Militia yesterday, takes the earliest opportunity of expressing his surprise and gratification in their improvement since he had last the pleasure of seeing them in the field. Major-General Whitelocke is satisfied that the General officer, under whose orders the regiment is about to serve, will place a proper value on the excellency of its discipline. Lieut.-Colonel Ravenshaw, having commanded the regiment nearly the whole of the last six months, is assuredly entitled to infinite praise for his zeal and exertions.

"THOMAS PRETYLER,

"Major of Brigade."

June. The Berkshire Militia marched into Weymouth and were encamped near the town. The King and the

Royal Family were, at this time, staying in the town.   His
Majesty seems to have taken great interest in the soldiers.
He ordered for himself a special uniform to wear during
his visit to Weymouth, and desired the Princesses to
adopt it.   It was scarlet, with narrow gold lace, the same
as worn by the Knights of the Garter.

July 22nd.   The King and the Royal Family reviewed
the Berkshire Militia near their encampment at Weymouth.
Eleven of the principal manœuvres agreeable to His
Majesty's regulations were selected for the day's performance, which they went through in such a superior style as
to gain the approbation of His Majesty, Prince Adolphus,
the Generals, and numerous Field officers present, who
pointedly expressed that they had never seen a regiment
execute those manœuvres better : " Very great praise is due
to Colonel Vansittart, who (so lately appointed to the
command of the regiment) has, by attention to his military
duty, gained so much knowledge of his profession as to
manœuvre his regiment and deliver his words of command
so remarkably correct.   He was well supported by every
individual officer in his regiment ; and to such perfection are
the men brought, that during the review not one false
motion was made."

After the review His Majesty and the Queen, and Royal
Family, with numerous nobility, partook of an elegant
*dejeuné*, given by the Colonel, in tents, in front of the Berkshire line of encampment.   Mrs. East and Miss Laura
Vansittart, with Colonel Vansittart, waited on the King
and Queen; while the Hon. L. Bouverie, Lieut.-Colonel
Ravenshaw, Major Pye, and Captain East were attendant
on the Princesses and the select party in the King's tent,
amongst whom were the Countesses of Radnor and Rosslyn,
Ladies C. Durham, M. Wynward, C. Enken, C. Bellairs,
Hon. Mrs. Grant, &c.   The Princesses, with their usual
affability and condescension, expressed to their attendants

the pleasure they had received in the very excellent review of " their County Regiment."

Colonel Vansittart appointed Reading as the headquarters of the regiment for assembling, prior to the disembodiment.

August. The Supplementary Militia were assembled in Reading, and thirty or forty men were selected to join the Regular Militia, the rest were given two days' pay and dismissed to their homes.

A court-martial was held on two of the York Hussars, who had deserted and cut a boat out of harbour with the intention of going to France, but by mistake they landed at Guernsey and were taken prisoners. Sentence was passed upon them, and they were shot on Bincombe Down, near Weymouth, June 30th, 1801. All the regiments, both in camp and barracks were drawn up. The Greys, the Rifle Corps, and the Staffordshire, Berkshire, and North Devonshire Militias. The men came on the ground in a mourning coach, attended by two priests. After marching along the front of the line they returned to the centre, where they spent about twenty minutes in prayer, and were then shot by a guard of twenty-four men ; they dropped instantly, and expired without a groan. The men appeared sensible of their awful situation, and were very penitent. The soldiers then wheeled in sections and marched by the bodies in slow time.

October. Lord Hobart, in a circular letter to all the Lords-Lieutenants in the kingdom, had, by the King's command, expressed His Majesty's " deep and lasting sense of their steady attachment to our established Constitution, and that loyalty, spirit, and perseverance, which have been manifested by the several corps of Yeomanry and Volunteers in every part of the kingdom. Farther, that they (the Lords-Lieutenants), at the next meeting of the corps, will, in His Majesty's name, thank them, and request that they will continue themselves in readiness for immediate services until the definite treaty is signed, as, till then, it is necessary that there should be no relaxation in the preparations which have been made for the general defence." This letter likewise directs the suspension of the measures ordered pursuant to the Act of the 38 George III. in the event of invasion.

December 11th. The Berkshire Militia, from Winchester, arrived at Reading two companies at a time. For nine years the regiment had been helping with the defence of the country, and had not been in its native county, and as they entered the town the men gave three hearty cheers to be home again. The following day they proceeded to Henley.

The whole time the regiment had been away on duty they had not a single complaint against them, but had everywhere gained the respect of the inhabitants at the various stations at which they had been quartered, greatly to the credit of both officers and men.

The Militia Bill for this year allowed the King to assemble the Militia for twenty-one days only, if he thought fit, it ordered one third to be drawn out annually. The men being drawn for ten years, and the substitutes for five years, to be renewed at the end if desired. For regiments 800 strong, one Colonel, two Lieutenant-Colonels, and three Majors were allowed. For regiments of 700, one Colonel, two Lieutenant-Colonels, and two Majors.

The Treaty with France was not definitely signed, and until that took place disembodiment was postponed. At first all officers and men were ordered to reassemble, previous to disembodiment, on March 10th, but owing to delays the date was altered to April 14th. All over Berkshire the news of the Proclamation of Peace was received with great rejoicing. An address was sent to His Majesty from the royal county. It was conveyed to town by the high sheriff and a large party of county gentlemen, the following of whom kissed the King's hand for the first time: Sir Joseph Andrews, Bart., Woodcock Croft, Esq. and Thomas Goodlake, Esq.

1802. The Treaty of Peace, called the Peace of Amiens, with France was signed, and the Annual Register states "that April 24th, 1802, the whole of the Militia and fencible regiments were disbanded, and the reduction of the regular troops will speedily take place."

The Berkshire Militia was at Weymouth. They were marched homewards and quartered at Henley, Maidenhead, Marlow, and Wycombe.

March. The 7th Dragoons had been for some time in Reading; they were marched out to make room for the Berkshire Militia. Three months previous the Militia had passed through Reading in divisions, but now after nine years embodiment, they were returning to the town. They were welcomed by the ringing of the church bells and the shouts of the vast crowds who collected to see them enter the town. Wherever they had been quartered during those nine years, the regiment had earned respect and approbation, and the Reading people were justly proud to note the fact.

On April 24th, the disembodiment took place. At the same time the Commander-in-Chief issued a circular letter, sent to all Militia agents, which was thought to be most considerate on his part, showing his anxiety for the welfare of those under his command.

"Adjutants (the only officers remaining in pay after the regiment is disembodied) are to take charge of all the men who are unable to march home, that they are to subsist them, and provide every medical assistance until they are able to march home, and that in the meantime their bills for what is wanted for that service for the monthly estimate are to be honoured by their agents." This was a vast improvement for the men's comfort, as previously they had been without assistance or relief of any kind.

# CHAPTER X.

N March 30th, 1803, the Berkshire Militia was again embodied at Reading, under Colonel Vansittart.

They were a very fine body of young men, but the muster was deficient. This, however, was remedied a week later, due notice having been given that all such as had been balloted for and had decamped before the constables had time to serve them with the notice, were quite as liable as those who were sworn and then deserted; in either case they would be punished as deserters and sent for foreign service, unless they surrendered themselves at once to the sergeants. Arms were delivered out to the soldiers, and so eager and attentive were both officers and men that they soon made great progress, and could compare with any corps in the country.

The regiment marched out from Reading to Ashford Barracks, where they were quartered with the Royal Surrey Militia. The Supplementary Militia of Berkshire was then called out for duty, to protect the county while the old Militia was away.

It was at this time that volunteer corps and local cavalry were raised in every division of the county. Many former Militia officers, who had resigned, now came forward and took commissions in these local regiments.

Among other regiments were the "Aldermaston Fencibles," in which my great-grandfather, William Thoyts, of Sulhamstead, held a commission.

One very patriotic individual wrote to the *Reading Mercury* to say that he had spent thirty-seven years of his life in the Berkshire Militia, and six years in the Marine service at the time of the Rebellion ; that, as he was only 79½ years of age, he still wished to serve his country in such a critical time ; he could make blunt (*sic*) cartridges and ball cartridges, and evidently had some idea of setting up a factory at Luckley House, Wokingham, whether he did or no the paragraph does not say, nor does it give the name of this worthy veteran.

About this time 180 of the Berkshire Militia volunteered into the Line.

The oldest records in the possession of the Regiment commence in this year, consisting of a Court Martial Book and a Register of Officers, both these books, curiously, were bought (so says a label on the covers) from a Nottingham stationer, which makes me think that they are copies made in 1812 from older papers or books of 1803.

1804.　Among the Militia correspondence at the Record Office is this letter from the Secretaries to the Right Hon. Charles Yorke, dated from the War Office, April 23rd :

"Gentlemen,—I have the honour to enclose, for the information of Mr. Yorke, a copy of a list transmitted to me. by command of His Royal Highness the Commander-in-Chief, of regiments of Militia which have received His Majesty's gracious permission to bear the appellation of Royal Regiments.

"I am, &c.,

"F. MOORE."

A list follows of regiments of Militia which received His Majesty's gracious permission to bear the appellation of Royal Regiments, and which claim as such the distinction of wearing blue facings. Berkshire is among the list which contains twelve regiments. Colonel Davis, in his history of "The West Surrey Militia," also quotes the above letter, but the list he gives of the Royal Regiments names thirty-two regiments. His extract must have come from a different document to the one copied for me from the Record Office, although it bears the same date.

The Duke of York, with paternal and becoming regard for the welfare of the soldiery (no less creditable to his feelings as a man than to his watchfulness of their health as Commander-in-Chief), caused to be circulated the following letter from the Horse Guards :

"Sir,—The Commander-in-Chief having observed with infinite regret the fatal effects the small-pox has, in seventy recent instances, produced in the Army, His Royal Highness apprehends that sufficient attention has not been paid to the order respecting the Vaccine inoculation, issued on November 18th, 1803. His Royal Highness therefore requests that you will recall the order to the recollection of officers commanding brigades and regiments; and that you will enjoin them to give it all possible effect, by explaining to the men the beneficial consequences resulting from the inoculation of the cow-pox, which has long been proved to the entire conviction and satisfaction of those who have had the best opportunities of observing the mild and rapid progress of this important discovery.

(Signed)       "HENRY CALVERT,

" Adjutant-General."

Before the Berkshire Militia left Ipswich, a grand review took place on Rushmere Heath by H.R.H. the Duke of York, who inspected the whole garrison, which then consisted of the Royal Berkshire, Royal East Middlesex, West Suffolk, Shropshire, and Herefordshire Militias and Artillery, commanded by Lieut.-General Lord Charles Fitzroy. The Cavalry being the 2nd or Royal North British Dragoons, 21st Light Dragoons, 7th Light Dragoons, and a detachment of Horse Artillery, commanded by Major-General Lord Paget. After the review the following orders were issued :

"Lieut.-General Lord Charles Fitzroy has the most sincere satisfaction in making known to Major-General Robinson and the two brigades who were yesterday reviewed, the very handsome manner in which His Royal Highness, the Commander-in-Chief, was pleased to express himself at the very high order in which the whole appeared, and at their great steadiness and attention in the field. If Lieut.-General Lord Charles Fitzroy may presume to add a few words of his own, he will consider this a fair opportunity of returning thanks to those commanding officers whose exertions, seconded by the officers and non-commissioned officers, and by the good discipline, steadiness and handiness of the men under their command, ensured the whole as well as each individual regiment, to appear with so much credit to themselves, as well as with real pride to the Lieut.-General commanding them."

1805. October 30th and 31st. The Berkshire Militia passed through Reading on their way from Ipswich to Taunton in Somersetshire. To welcome them the bells of the three Reading churches rang gaily, and hundreds of the inhabitants of the town assembled to see the regiment, which had been away on duty for two years.

1807. The War Office Papers give among the Militia uniforms of 1807, the Berkshire Militia's, at Reading :

Staff-Sergeants' clothing : Sergeant's and Drum-Major's coats : scarlet with blue facings, silver Bias lace and fringe for both, and plated buttons.

Sergeants' clothing : Coats scarlet with blue facings, ⅜ in white lace, with the shoulder straps sewed in the seam of sleeve, looping jews harp ten by twos as pattern drawn.

Privates' clothing : Coats red with blue facings, with a white worsted lace with a blue edge, threads blue inwards looping the same as sergeants' ten by twos.

Drummers' clothing : Coat red with blue facings, with fringe on wings and under the darts of sleeve, white worsted and broad lace ⅞-in. over the seams of body and sleeves of

LIEUTENANT'S. _____ [Pale Yellow.]

PRIVATE'S. _____ [Paler Yellow with a Dark Blue edge.]

DRUMMER'S LACE [Yellow ground with Dark Blue edge and Pattern.]

DRUMMER'S LACE. [Yellow ground with Dark Blue edges and Pattern.]

PRIVATE'S COAT.

DRUMMER'S COAT.

DRUMMER'S SLEEVE.

coat, pocket frame and two strips leading from the pocket frame to the arm holes and down the front edge from the collar to the bottom of turnbacks, narrow and broad to the hips and cuffs, narrow lace ½-in. for looping the coat and collar and wings with looped strips as pattern drawn, looping ten square loops by twos as pattern drawn.

Button Maker: Mr. Shaw, London.

### LOCAL MILITIA.

1809. Three regiments of Local Militia raised in Berkshire: first regiment, ten companies, F. Page, Esq., Lieut.-Colonel, Commandant; second regiment, eight companies, Henry John Kearney, Esq., Lieut.-Colonel, Commandant; third or Royal regiment, seven companies, George, Marquis of Blandford, Lieut.-Colonel, Commandant.

The Local Militia was established under Statute 48, George III., cap. 3, which received Royal Assent June 30th, 1808, and their services were continued to the end of the then war, namely, May 1st, 1815, by Statute 55, George III., cap. 76, which received Royal Assent on June 14th, 1815, four days before the Battle of Waterloo.

They were called out annually in the spring for training, and took the place of the regular Militia during its absence on service. They were looked upon as a sub-division of the Berkshire Militia, and were always spoken of as the Local or New Militia, and were distinct from the Volunteers of a later date which had no connection with the Militia.

June 25th. While the Local Militia were exercising in the Forbury, on being refused their marching guinea previous to their dismissal the next day, several companies laid down their arms, to which conduct it was afterwards asserted they had been incited by some of the Volunteers imprudently urging them on and promising to stand by them. How far this assertion is founded on fact we know not, but certain it is, the officers of the Volunteers were by no means impli-

M

cated in the charge, neither could it be expected that they should be answerable for the conduct of their men while off duty.  Notwithstanding this, however, on complaint being made at the War Office by the Commander of the Local Militia, of the supposed misconduct of the Regiment of Reading Volunteers, they were, in the July following, dismissed from His Majesty's service without the smallest remuneration and without even a compliment being paid them for their meritorious conduct in coming forward in the hour of danger to serve their country, almost wholly at their own expense.  *(Man's History of Reading).*

1811.      The Militia force in England, on June 25th, was 77,424 men.

The riot of frame breakers, known as Luddites, began at Nottingham in the middle of November.  The wholesale hosiers for some time had been curtailing their hands, and much distress and discontent ensued.  Certain large frames had been introduced which effected a considerable saving in manual labour, and this incensed the workpeople ; they commenced breaking all these frames, and though the master weavers armed themselves and barricaded their houses, the mob was too strong for them, and continued the outrages.  The Military were called out, consisting of thirty men of dismounted Dragoons, the *posse comitatus* or Militia, the 1st and 2nd Local Militia, a special messenger being sent to town to ask for further aid to quell the riots.

Besides the troops named, two troops of Volunteer Cavalry and a detachment of the Queen's Bays took up their quarters in the town.  At this time the Nottingham Militia, called the "Sherwood Foresters," were taking their turn of two years duty in Ireland.  The presence of so many soldiers at first overawed the frame-breakers, and tranquility was restored, but only for a brief interval.  The Berkshire Militia arrived at Nottingham soon after the outbreak, and continued there during the rioting.

1812.  Riot and frame-breaking continued all through January and spread to Derbyshire. Night attacks occurred on peaceable individuals, so that watch had to be kept. The Corporation of Nottingham took measures for protection, but this only made the frame-breakers more cautious. They fired a gun as a signal, and dispersed before they could be captured by the watchers. However, captures were made of suspected persons; among others, a celebrated pedestrian, well known as a deserter, who had long been a terror to the neighbourhood.

An elderly woman, wife of a person who owned seven frames broken at Basford, swore to several persons concerned in the outrage, two of whom were committed. So much indignation was excited against her in consequence among the stocking makers of Basford, that it was judged expedient to remove the family and their furniture, escorted by the military at Nottingham; the Sherwood Foresters were at this time in Ireland and the Berkshire Militia quartered at Nottingham, so this duty must have fallen to them.

Forty men, disguised in various ways and armed with pistols, etc., went to Mr. Benson's house, placing sentinels at the neighbours' doors and the avenues leading to it; about eight of them entered the house, driving the family into the pantry, with threats of immediate death if they created the least alarm. The rioters then went to the workshop and demolished the eight frames.

Two days later eight men, carefully disguised, went to Mr. Noble's house, at New Radford; they destroyed four warp lace frames, because they were making what was called a two-course hole. In vain the owner, Mr. Noble, told them he was receiving 8d. a yard over the standard price. "It was not the price but the sort of net they objected to." He narrowly escaped being struck down as one of the ruffians cut at the frame with a sword. Mrs. Noble was struck

M 2

on the head with the butt end of a pistol ; a neighbour entering at the time tried with Mr. Noble to secure the man, but he and his companions decamped.   Similar attacks and other outrages took place in and around the town.

A picket of one hundred men paraded the streets of Nottingham in separate parties, headed by the Civil authorities, every night, ready to read the Riot Act.

This, however, was of no avail, for more frames were still broken.   In one night alone, thirty or forty were destroyed, and by this time the Luddite riots had spread to Yorkshire.   It was rendered more alarming by the secrecy with which the outrages were planned and carried out.   In most of the villages where so many frames had been broken, parties of the military were stationed, but their exertions had been inadequate towards the apprehension of the offenders.   The rebels assembled and dispersed in a moment, they were marshalled and disciplined like a regular army, and commanded by one particular leader, called Ned Lud, under whose banner they swore to conquer or die.

General Hawker went to Bulwell, a manufacturing village about eight miles from Nottingham, with a strong party of the Berkshire Militia and two officers to quell a serious disturbance there.   Two other regiments of infantry received orders to march forthwith to Nottingham, as the military in the town were not strong enough for the public security of the country.   Some Bow Street officers were also sent from London.

The constant parading of the military, their movements in various directions both by night and day, gave the appearance of real warfare.

At Lenton, within a few hundred yards of the Barracks, twenty frames were destroyed in one night.   After doing this the Luddites crossed the River Trent, and broke fourteen frames at Ruddington and twenty at Clifton, leaving only two whole frames in the latter town. ' An express was sent to Nottingham for a troop of Hussars, who went with all possible speed, and as many of the Bunney Troop of Yeomanry as could be collected (they being in the neighbour-hood of the scene of action) were immediately mounted. One party pursued the depredators, while others seized all the passes over the Trent for a space of four miles, under a full persuasion that the Luddites could not escape; but such was the generalship of their leader, that they seized a boat which nobody else had thought of, and re-passed the river in two divisions in perfect safety and escaped.

That same night a frame was broken at Bulwell, while a sergeant and six men of the Berkshire Militia were employed to watch it. The parties exchanged shots several times, but it is not known that anyone was wounded, though one of the Luddites lost his shoe and a hammer.

The next night, Sunday, more frames were broken. At Basford one case was most daring. Three soldiers were in a house protecting three frames belonging to William Barnes. A party of Luddites entered the house and immediately confined the soldiers, and while two of the party stood sentry at the door with the soldiers' muskets, others demolished the frames. After the mischief was done, the muskets were discharged and the soldiers liberated, the depredators wishing them good-night.

Three regiments of soldiers, the local police, and the Bow Street officers were powerless to stop the mischief. Many people were taken up as suspected frame-breakers; but no evidence could be got, nor was it believed any of those appre-hended were frame-breakers. Four of these prisoners were brought into Nottingham with great parade, by three several parties of military and civil officers. Two of them were persons who had had frames broken in their own houses, and another was a well-known maniac named Waplington, a pauper, who for years had been in the habit of wandering about. It excited much laughter to see a Bow Street officer with this poor creature confined in a cart by his side, driving furiously along the streets, guarded by half-a-score of Hussars.   It is supposed the maniac was caught on one of his wandering excursions and refused to give any account of himself.

In one outrage the Luddites, about twenty-four in number, disguised themselves in soldiers' great coats, with their leader carrying a large staff, so that the people thought they were the nightly picket capturing some rioters to take them to prison.

Some of the Scots Greys from Leeds were ordered to Huddersfield, the riots having spread there; and a squadron of cavalry were sent from Sheffield to take their place at Leeds, until a squadron of the 2nd Dragoons from York Barracks arrived at Huddersfield, when the Scots Greys were sent back to Leeds again.

At Nottingham the rioters gave up frame-breaking (perhaps because there were none left to break) for worse atrocities.   They placarded the streets one night, offering a reward for Mr. Wilson, the Mayor, "dead or alive," because he had offered a reward of £590 from the Corporation of Nottingham, for the discovery of the assassin who shot Mr. Trentham.   The rioters grew so bold, they actually met in small parties in the streets of Nottingham in broad daylight and triumphantly talked over their nightly depredations.   Reinforcements had further to be sent, consisting of two rifle companies of the North York Regiment.

April.   From Nottingham, in the spring of the year, the Berkshire Militia were sent to Lancashire to help suppress the riots which had spread to those parts.  On April 30th, 300 of the regiment, under Lieutenant-Colonel Ravenshaw, passed through Derby, on their way from Nottingham to Manchester.   They travelled in waggons, and a tedious journey it must have been, for the direct route leads across the hills of Derbyshire, through Buxton to Macclesfield, along the Manchester high-road, a high-road in every sense of the word. From Manchester they were sent to Liverpool, and thence to Preston and Blackburn. The riots were in full force over all that district.

From Lancashire they were ordered to the South of England, and arrived at Sommerton expecting to be sent to Plymouth or the extreme West of England, where disturbances had arisen in the shape of corn and potato riots, especially in the neighbourhood of Plymouth. The route for the regiment was altered, and they were sent to Bideford. The principal corn crops were, however, harvested without much injury, and military force was not required. It was not until December 28th that the Berkshire Militia left Bideford to take up their quarters in Mill Bay barracks at Plymouth, remaining in the neighbourhood of that town during the winter and spring of 1813.

1813.  Two French officers on parole fought a duel at Reading, not far from the "New Inn," on the Oxford Road. Unable to procure pistols, they used fowling pieces at fifty paces, firing alternately. The first shot was conclusive, the ball entered the unfortunate man's neck. He was conveyed back to his lodging in a postchaise, where a surgeon dressed his wound, which was said not to be dangerous. The French prisoners about this period were let out on parole, which many broke; some succeeded in escaping, but most were recaptured. The return of prisoners of War in Great Britain then gave: French, 52,649; Dutch, 1,868; total, 54,517.

A young man at Horsham was committed to the gaol under the following circumstances: He had formerly lived near Rye, then went to London, and lived most expensively till he had spent all his money; after that, he set his wits to work. It occurred to him that, knowing the coasts well, he might make a considerable sum by conducting the escape of French officers at large on parole. He contrived to get

acquainted with a French Colonel and Major at Reading,
who, glad of the opportunity, agreed to give him 300 guineas
if he assisted them to escape; £150 was paid at once, and
the rest was to be handed over as soon as they were on
board the boat. The party set out in a postchaise, and
arrived without interruption within a few miles of Hastings,
where they were observed by an exciseman, who, from
their appearance, suspected them to be escaped prisoners;
although the landlord of the inn told him they were German
officers on their way to Bexhill to join their regiment. The
exciseman got a party of military and seized the French
officers in their beds. They made no attempt at disguise,
and submitted " in a handsome manner," the young man own-
ing that the 300 guineas tempted him in the exhausted and
desperate state of his finances. The French officers were
" properly disposed of." Poor men, doubtless they were sent
back again to prison and misery, to brood over their bad
luck and the nearness of their escape.

The next important event in the Berkshire Militia annals
was the departure from Plymouth for Ireland. One
authority says they sailed on April 29th, another gives the
day as May 7th. Perhaps both are correct, as the regiment
probably required more than one vessel for its embarkation.
The voyage took several days; but the Cove of Cork was
reached on May 12th, the regiment arriving there in perfect
health. They at once marched to Middleton and from
thence to Athlone, where they were quartered until the
autumn.

June 10th. A sergeant, a drummer and two privates,
with a party of thirty-eight recruits, were ordered to leave
Reading, and travel as quickly as possible, to join the
regiment in Ireland.

August 16th. The Royal Berkshire Militia left Athlone
for Galway. Whether they remained there for winter
quarters does not appear clear, no information seems forth-

coming on the subject. That they saw service in Ireland is certain, but where or when, history does not relate. While the Berkshire Militia were serving in Ireland, their noble example was followed by the Local Militia, who volunteered for foreign service in December; but they were too late to be accepted as a regiment, though many of them were passed on into the Regulars.

1814. September. The Berkshire Militia were at Tuam, which they left for Newry and the North of Ireland on their way homewards. All regiments being ordered back to their own counties, to be disembodied as soon as convenient; for Peace had been formally proclaimed, with the customary rejoicings, in the previous month of July. The War was over, as far as home defence was concerned.

*The General Advertiser* or *Limerick Gazette*, of Tuesday, September 27th, 1814, gives the following paragraph:

"Three regiments of English Militia have received orders of march for Newry, to embark for England : The Cambridge, from Lifford, to arrive the 26th and 28th instant; the South Lincoln, from Londonderry, to arrive the 27th; and the Berkshire, from Tuam, to arrive on the 29th and 30th."

From another Irish newspaper of the same date, *Saunders's News-Letter and Daily Advertiser*, Friday, October 7th, 1814, the extract shews that in Ireland, as elsewhere, our Militia won respect and esteem. In this case it must be borne in mind that Ireland was, at that time, a hostile country, so this praise has additional value.

"On Monday, the South Lincoln, the Berkshire, and Carmarthen Regiments of Militia marched out of this town, for Warrenpoint, to embark on board the transports for England. It is but justice to say of the five regiments, for some days billeted on this town, viz., the Cambridge, South Lincoln, Berkshire, Carmarthen, and Monaghan, amounting, including women and children, to between four and five thousand persons, that their conduct has been most correct and exemplary. The English Militias, returning to their country, carry with them the good wishes of the inhabitants of Ireland; and we trust and hope our countrymen, on leaving the sister kingdom, are equally beloved and regretted. The Cumberland Militia are expected here in a few days, to embark for England. The Monaghan, it is said, are to go into barracks."

*Newry Telegraph.*

Two ships were provided to convey the Berkshire Militia across the Channel, one was an old steady-going vessel, the other a new, fast sailing one; the first arrived after a passage of three days, but the other encountered a gale and was tossing about for nine days before she reached Liverpool, and it was feared she had foundered.

So soon as the regiment landed at Liverpool, it was expected they would be ordered at once to Reading. This was done, and the order was issued for them to start on October 18th and 19th on the long march home, where they were timed to arrive November 3rd and 4th; but this order was cancelled at the last moment, disembodiment being suspended, and the Berkshire Militia remained at Liverpool until further notice.

The return to their native county took three weeks to accomplish. By what route they travelled is not known, as the marching orders for those years are missing. Finally, however they marched into Reading in one division, in September, 1815. where they remained until March 14th, 1816, when the final disembodiment took place. The Militiamen who fought at Waterloo must have volunteered to the Line before their own regiment was disembodied.

About this period, soldiers' allowance on march was much reduced. The innkeepers up to 1814 were allowed 1s. per diem, but in 1816 it was reduced to 10d. per man.

Alderman Darter gives the following anecdote of Colonel Ravenshaw :—

There was in the regiment "a young and very efficient officer, who was not on very cordial terms with his brother officers. The Colonel seems to have noticed this, and subsequently ascertained that the cause arose from the officer in question having relatives engaged in trade, and this fact induced them to 'cut him.' This state of things led the Colonel to pay marked attention to him, occasionally walking with him on parade before the men fell in, but he did not in like manner notice the other officers." This had the desired effect, and afterwards the officers became friends.

As soon as the Militia returned to Berkshire they were disbanded, and as at this time the war on the Continent was still engaging our troops, recruiting parties were sent all over England enlisting for the Regulars. Some must have gone to the Line directly they landed from Ireland. To Reading came the Blues and the 7th Hussars for this purpose, and as the Berkshire Militia were just paid off, numbers enlisted at once. Before Waterloo, there was not time to clothe them in the regimentals of their new regiments so they fought in their Berkshire Militia jackets, one of which was preserved in the museum at Waterloo, and was seen there by Alderman Darter; he names several who were in the Battle of Waterloo, and he also gives this interesting account of the old Militia Band :—

"When the regiment came home in 1815, it was finally stationed at Reading. After the conclusion of the war the regiment was disembodied and the band broken up, excepting what is called the drum and fife band, and the staff, which were retained, and were actively engaged whenever the Militia was called out. On ordinary occasions both staff and band were drilled either in the Forbury or in a meadow adjacent to Captain Perron's residence, by an excellent non-commissioned officer, Sergeant-Major Preston. The drum and fife band was the best for its number that I ever heard, and it was a great treat to go to the Forbury on a Sunday morning in those days and see the staff march to St. Lawrence's Church with the band at their head playing ' Hark, the bonnie

Christ Church bells,' and after service return, marching to a quick step.  My memory dwells on the whole scene as it was at that period, before modern requirements interfered with the neighbourhood and intercepted from our view some of the beautiful scenery of Oxfordshire.  To revert again to the band, there were some excellent musicians in it, some of whom will be even now remembered.  For instance, the two brothers of the name of Kates, the elder of whom is mentioned in Mr. Binfield's programme of the musical festival in the year 1819.  I must not omit to mention the name of Penny.  He was the very best player on the small drum that I ever heard, although the mention of this may seem too trivial to be worth recording (except to Reading men), I have seen little groups of people, and amongst them soldiers of other regiments, wait in London Street to hear him beat the usual reveillé.  The music for this small band was arranged by the late master, Mr. Smith, who was with the Berkshire Militia in Ireland : it was unique of its kind, being before keys were applied to bugles, French horns, etc., the half tones being produced by inserting a hand into the bell of the instrument.  The music in question was arranged for flutes with a tenor drum, which was tuned, and beaten by a small fellow of the name of Cray ; also a large drum and a novel accompaniment for first and second bugles, played by Edward Blagrave and George Adams, and what they had to do was most effective.  I particularly remember a quick step in C, with three parts of eight bars each, which I picked up from hearing it often, so that I could at any moment write it, although seventy-two years have elapsed since I first heard it.

"All the men to whom I here refer have passed away with the exception of one.  He is, I believe, about eighty-three years of age, and was a boy on board the transport ship which brought part of the regiment from Ireland, and was eight days making the passage to Liverpool.  All the men, in both band and staff, were, with few exceptions, natives of Reading, and in the prime of life, but to quote an old ballad

"They will never march again."

"During the incumbency of Dr. Wise it was customary on New Year's Eve for the ringers of St. Lawrence's Parish to ring a few peals of changes and leave the bells up on their stays, and some short time before midnight to return. At the same time the Militia band assembled at the upper end of London Street and all was still until the moment St Lawrence's clock began to strike twelve. Then off went the merry peal of eight bells, and at the same moment three loud strokes of the big drum led off the Berkshire band down London Street to the Market Place, and from there through a portion of the town. Seventy-two years have elapsed since I first experienced the magic effect of this music of the band and the merry peal of St. Lawrence's bells breaking out in the stillness of midnight, suggesting that the old year had passed away and welcoming the dawn of its successor. After a short interval the old watchman, Norcroft, went up London Street, calling out ' Past twelve and a starlight mornin' ! "

The extract above is worth quoting, as Alderman Darter's "Reminiscences" were only privately published and are already scarce. The Band as above described appears to have consisted of a big drum, a tenor drum, first and second bugles, flutes and horns. One of the horns was called a serpent, from its twisted body. Doubtless the "March" mentioned, is the old one, written for the regiment in 1792, now quite forgotten, but which, to my great joy, I am able to restore again to the regiment.

1816. Now-a-days, when railways are universal and folks grumble at being a dozen miles from a railway station, it is hard to realise the time when everything and everybody journeyed by road.

. In July of the above year, the colliers and iron-workers of Bilston, being out of work, determined to go in person and represent their case to the Prince Regent, thinking he could order them employment; for they prided themselves they were willing to work, if work could be found them. They advanced as far as Maidenhead Thicket, when Mr. Birnie, the Bow Street Magistrate, who was accompanied by several special constables and a party of military, met them.

The colliers had with them a waggon weighing over two tons; it was drawn by forty-one men, and their leader rode on horseback beside it. The men refused to sell the coals but gave them up, whereupon they were made a handsome present, and the coal was distributed to the poor of Maidenhead. The colliers behaved so well that, after gifts of money had been presented to them, they were sent back to Bilston.

## LIST OF OFFICERS OF MILITIA, 1817.

ROYAL BERKSHIRE MILITIA, DISEMBODIED MARCH 14TH, 1816.

HEADQUARTERS, READING.

| Rank. | Name. | Date. |
|---|---|---|
| Colonel - - | Thomas William Ravenshaw - | Dec. 9th, 1812. |
| Lieut.-Col. - | William Viscount Folkestone - | ,, ,, |
| Major - - - | John Blagrave Pococke- - - | Nov. 19th, 1810. |
| Captains - - | Lionel Charles Hervey - - - | April 5th, 1805. |
| | John Sturges - - - - - - | June 5th. |
| | Samuel John Bever - - - - | Dec. 12th, 1807. |
| | James Woodhouse - - - - | Dec. 16th, 1808. |
| | Charles Velley - - - - - | Nov. 28th, 1809. |
| | James Goddard Doran - - - | Nov. 30th, 1815. |
| Lieutenants - | Henry Gilbert Stephens - - | Dec. 7th, 1793. |
| | Surgeon-Mate Stephen Judd - | July 18th, 1803. |
| | William Coles - - - - - | June 21st, 1809. |
| | John Austin - - - - - - | April 23rd, 1813. |
| | John Norris - - - - - | April 29th, 1814. |
| | John Iman Davenport - - - | April 30th. |
| | James Hance - - - - - | Oct. 15th. |
| | John Parker - - - - - | Oct. 16th. |
| | William Roe - - - - - | Feb. 2nd, 1815. |
| | William Phillips - - - - | June 26th. |
| Ensigns - - | Surgeon-Mate William Stratton | Jan. 19th, 1813. |
| | Paymaster Edmond Slocock - | Sept. 9th, 1815. |
| | Benjamin Hawkins - - - | ,, ,, |
| | James Ince - - - - - | Dec. 21st. |

| Rank. | Name. | Date. |
|---|---|---|
| Adjutant - - | Captain Edward Purvis - - - | Jan. 26th, 1813. |
| Qr.-Master - | Lieutenant George Guy - - - | Mar. 19th, 1807. |
| Surgeon - - | Charles Greenhead - - - - | Nov. 13th, 1814. |

### 1820.

Lieut.-Colonel John Blagrave, July 14th, 1817, in the room of
Lieut.-Colonel William Viscount Folkestone.

Lieuts. -
{
Henry Gilbert Stephens } Disappear from the Army
John Austin - - - - } List after 1820.
Benjamin Hawkins, appointed Sept. 26th, 1818.
}

Ensigns -
{ William Stratton - - - } Left after 1825.
Benjamin Hawkins - - }

Adjutant, Quarter-Master and Surgeon, remain.

### 1825.

Captain Charles Bacon, appointed August 8th, 1822.
Ensign James Ince, left after 1825.
All other officers remain as in 1820.

## THE ARMY LIST, MONTHY REPORT, 1826.

ROYAL BERKSHIRE MILITIA, READING.

| Rank. | Name. |
|---|---|
| Colonel - - - - - - - - - - | T. W. Ravenshaw. |
| Adjutant - - - - - - - - - | Captain Purvis. |
| Paymaster - - - - - - - - - | E. Slocock. |
| Quarter-Master - - - - - - - | Lieut Guy. |
| Surgeon - - - - - - - - - - | C. Greenhead, M.D. |

1827 and 1828, no change.

Paymaster - - E. Slocock - - - - }
Qr-Master - - Lieutenant Guy - - - } Disappear, July, 1829
Surgeon - - C. Greenhead, M.D. - }

1830 to 1841, no change.

Colonel T. W. Ravenshaw disappears from the List, Sept., 1842.
Colonel John Blagrave appears on the List, Nov., 1842.

Adjutants - -
{ Captain Purvis, disappears March, 1846.
,, Sherson, appears ,, ,,
}

1847, 1848 and 1849, no change.

In the autumn of 1830 the labourers of the southern
counties assembled in crowds, demanding more wages.
They destroyed mills, broke machinery, and set fire to stacks
and farm buildings.  The old methods of farming were
giving place to the new, machinery was doing the work
hands had hitherto had to perform ; and as in 1811 the
weavers had tried to prevent the use of larger frames in
their trade, so now the agricultural labourers rose to struggle
against improvements in farm implements, chiefly the use of
the thrashing machine instead of the old hand flail, excited
their wrath.  Wiltshire, Hampshire, Berkshire, and Kent were
foremost in these riots.  Special commissions were called at
Salisbury and Ely to try the rioters, many of their ringleaders were hanged.  Soldiers were called out to protect
life and property.  The rioting in Berkshire was easily
suppressed.  It was never serious, being checked before it
became so.  Three men were condemned, but only one hung,
several others being transported.  In Berkshire the Militia
was ordered to be embodied.  Some of the officers who had
served in the old war were available, and commissions were
given to fill vacancies, but the danger passed without need of
the soldiers.  The new officers thus commissioned never
served, nor even possessed or wore uniforms.  Among
these were my grand-father, M. G. Thoyts, A. C.
Cobham, of Shinfield, and James Wheble of Bulmershe,
and as the regiment was not required for duty these
officers soon resigned their commissions.  The Army List
continued to give a list of the Berkshire Militia officers
annually.

# CHAPTER XI.

## CHANGES AND IMPROVEMENTS.—1852-1872.

THE Militia has very little worth recording until the year 1852, when all over England the regiments were reorganised, and two-thirds of the strength of the Berkshire Militia was called out for training, the whole regiment being raised the following year. Balloting was not then resorted to, nor has it been employed since, though if necessary it could be employed.

The Duke of Wellington was strongly in favour of the reorganisation of the Militia, his speech on the subject is well-known, but he saw clearly it required different organisation to render it efficient, and he objected to the rivalry existing betcen it and the Line Regiments. He, like all officers of the present day, wished to make the Militia a part of the Regular Army, equivalent to it in every particular.

Pressing for the Army had always rendered soldiering unpopular; the system of substitutes, and the large sums paid for them, had been one of the abuses of Militia regiments in the old war time. Voluntary enlistment did away with this. The quota in 1852 was raised to 777 men, yet there was little or no difficulty in getting plenty of men to enlist.

"Come on young men," said the Recruiting Sergeant, as he walked round the Reading Fairground in the Forbury. "Come on and enlist, one volunteer is worth ten pressed men."

The first thing was to see what remained of the old Militia, this the Adjutant, Captain Alexander Nowell Sherson (from the 72nd Regiment), at once proceeded to do. First he made enquiries as to the existence and where-

N

abouts of previous officers; of these few could be found, some had disappeared, many were dead, others were old and infirm.

The nominal return of officers gave:—Colonel: John Blagrave; Lieut.-Colonel: Charles Bacon; Major: John Leveson Gower; Five Captains, namely: Henry Greenway, James Joseph Wheble, Hon. E. Pleydell Bouverie, Edward Blagrave, and Henry Pole; Six Lieutenants: William Coles, John Norris, James Hance, John Parker, William Roe, and Henry Ince; Four Ensigns: Arthur Deane, James Winckworth, George Thomas Coleman, and Sir George Philip Lee, Bart.; Surgeon: Charles Greenhead.

Out of these only the Colonel, Lieut.-Colonel, Major, one Captain, one Ensign, and the Doctor, were found to be effective, so fresh Commissions were issued to complete the establishment, the Regiment not being fully officered until its embodiment in 1855. All the regiment possessed was found in the gaol, namely, twelve old muskets and two old drums (one, with the head knocked out, was used to store old papers in). I suppose the "papers" referred to, consisted of the two Court Martial Books of 1803-1815, and the Register of Officers of the Regiment, 1803, now in the Orderly Room, for they are the only old records existing.

The first training took place in Reading, November 11th to December 1st, twenty-one days. They still used the old ground in the Forbury, which had been the drill ground as far back as the days of Charles I., now it is built over and all traces of the "Garrison" have disappeared. Every other available piece of ground in the town was utilised, each company being divided into squads and drilled by its own officers (all of whom had previously joined some Line regiment for instruction), assisted by the sergeants on the Staff and such others as could be obtained from the Regular Army as drill instructors.

The companies were distinguished by numbers from one to eight: No. 1 being the Grenadiers, and No. 8 the Light Company. These were ordered to be selected from the rest of the regiment in the spring or summer of 1855, the Grenadiers being chosen for height, and the Light Company for general activity.

1853. October 20th to November 16th. The whole regiment assembled at Reading, twenty-eight days instead of twenty-one days being allowed. It was a bad time of year to choose for out-door work, and the weather proved unfavourable, cold and stormy. The Mayor of Reading and the inhabitants expressed their appreciation of the soldiers' peaceable conduct, no light praise at the time when the men were billeted in the public houses all over the town, and temperance was not so fashionable as now-a-days, yet the regiment kept up its reputation for good conduct. Non-commissioned officers and privates of the Scots Fusilier Guards had assisted in the drill, and good progress had been made in the short time they were out in spite of all disadvantages.

1854. In this year they assembled in May, again for twenty-eight days. By this time the Berkshire Militia had nearly regained its former state of efficiency.

An old-fashioned house with a garden was rented in Merchant Place below Friar Street, and here four or five of the officers lodged. The house belonged to the British Champagne Company, then managed by a man named Whittaker, an excellent musician.

The Mess was established and fitted out in 1854 with the necessary plate, kitchen utensils, etc. A warrant for embodiment was issued on November 27th, and on January

1855. 1st the regiment assembled at Reading. As in former generations the old Militia had always been first and foremost in war times, the old traditions were kept up by both officers and men volunteering for active service,

N 2

all the men cared to ask was whether their officers were going with them. Their offer to serve in the Crimea was accepted by the authorities, and the regiment received orders to proceed to the Ionian Islands, an important garrison at that period.

The precedency of the Militia Regiments was rearranged. In 1778, the Berkshire Militia had been 30th on the list, in 1833 its number was 7th, and so it still remained in 1852 and was confirmed in 1855. This again was altered when the Territorial System was established ; the 49th (Hereford-shire) Regiment was joined to the 66th (Berkshire) Regiment the latter then becoming the 2nd Battalion of the new regiment, the old Berkshire Militia becoming the 3rd Battalion. The latter were deprived of their title of Royal, given them by George III., 1804, a matter of serious dis-satisfaction to the whole regiment, who were extremely proud of a distinction not then possessed by either of the Line Battalions. It was, however, conferred on the regiment in 1886, for bravery of its 1st Battalion in the Soudan War.

The Berkshire Militia remained in billets in Reading from January 1st until it went to Portsmouth, *en route* for the Mediterranean, in September. They had volunteered their whole strength to the Line, or more, and yet they were as strong as ever again when they sailed to Corfu. It is said 700 men of the Berkshire Militia had by that time joined the Regulars.

Another fact still more to the honour of the regiment was especially noted by the Commander-in-Chief, namely, "that the Royal Berkshire Militia was almost the only regiment of Militia in which there had never occurred a Court-martial." Looking back, the old regiment had always been first and foremost in war times, eager to serve, indeed, many men were punished in the old days for deserting to join the Regulars for war service. In 1799, it responded to the appeal so readily that only half the men

THE COLOURS, MAGDALEN 1914

could be accepted, and, after the regiment was disembodied, numbers of its soldiers enlisted and fought on the Continent. A whole troop of the 7th Hussars at Waterloo were said to be composed of Berkshire Militiamen, and many joined the Blues and fought at Waterloo. Few counties in England furnish more recruits to the Army than Berkshire does, and if the regiment is below its strength the fault lies entirely with the recruiting arrangements.

Eager to testify their appreciation of their county regiment the ladies of Berkshire freely subscribed a large sum of money, part of which was expended in handsome colours, worked by the ladies themselves. The first, the Queen's colour with the Union, and the name of the regiment surmounted by a crown. The other, a rich royal blue silk bearing the regimental crest, a stag guardant, and the motto " Pro aris et focis." This crest puzzles me, I find that one of the crests borne by the Earl of Lovelace is a stag ; his ancestor was Lord-Lieutenant of Berkshire in the reign of Charles II. Yet after that date I know the regiment bore on their colours and on their drums the arms of the Duke of St. Albans, *i.e.*, the arms of England with the bar sinister, and of the colours applied for in 1762, I can find no details.

These colours were presented July 31st, and at the same time a centre-piece for the mess table in silver, representing the badge of the county, was given to the regiment from the surplus of the subscription raised for the colours. The afternoon of July 31st proved cloudy and wet, an unfortunate day for any out-of-door entertainment, but nevertheless the town of Reading was all excitement at the unusual ceremony about to take place. The spot chosen for the presentation was a large meadow near the Thames, approached from Vastern Street. It had been used as a parade ground by the regiment. In the centre of the meadow a dais had been erected for the Lord-Lieutenant and visitors. The ground was kept by the Hungerford

**Yeomanry** Cavalry under Captain Willes, of Goodrest House, and Captain Seymour, who had no easy task to manage the large crowd collected, especially at the narrow entrance to the field.

Admission was regulated by tickets, which had been given away by the officers of the regiment.   By two o'clock carriages began to stream down Vastern Street to the ground, while the foot people came also through the King's Mead and the Forbury in crowds.   Some thousands were assembled.   The air was thick and dark, rain began to fall, and any ceremony seemed hopeless.   At 2.30 the Lord-Lieutenant and the Countess of Abingdon drove on the ground in a carriage with the young Lord Norreys and Mrs. Blagrave.   They were escorted by a guard of honour of the Hungerford Yeomanry Cavalry.

The weather began somewhat to mend, and the regiment was drawn up three sides of a square.   The new colours were placed upon a tier of drums.   Since the regiment had been embodied the old colours had not been exhibited in public, as they were worn out, dingy with age and almost threadbare, a contrast to the glittering uniforms of the officers and men, (so says the newspaper report, but no one can remember ever seeing the old colours, nor is it known what became of them).

Sunshine succeeded the storm, and the ceremony commenced, the Lord-Lieutenant, Colonel Blagrave, the Rev. S. W. Yates, Vicar of St. Mary's, and the two ensigns, Cox and W. Stares, forming a group beside the colours.

The Rev. S. W. Yates had written the prayer of conse-cration, specially for the occasion, which he read with a clear, impressive voice, the Lord's Prayer was then read, after which the Countess of Abingdon unfurled the Queen's colour, presenting it to Ensign Cox, and the other to Ensign Stares, she spoke a few suitable words.

The regiment then formed up, marching in slow time, escorting the colours, with a guard of honour consisting of Captain Sir Paul Hunter, Lieutenant Thompson, and Ensign Tebbott. After this was over the Countess of Abingdon made a short speech, which was replied to by Colonel Blagrave, and another speech was made by the Lord-Lieutenant, Lord Abingdon. After this the ceremony was concluded.

In the evening a splendid banquet was given in the Town Hall by the officers of the regiment; covers were laid for four hundred. All the county neighbours for miles round were present, and several officers of the 94th Regiment. The evening was finished by a grand ball in the Town Hall, the liberality of the officers of the Berkshire Militia and their attention to their guests made a great impression on all who were present.

Before going out to garrison an important fortress like Corfu, a little practice at Sentry work was deemed advisable for the Berkshire recruits and this the Adjutant ordered. Anxious to carry out the order, Captain Slocock took one of his men to the towing path and stationed him there not far from a post, giving the order that "if any ships or boats came along, the sentry was to challenge them, and if they did not reply, to fire." The Adjutant walking that way, came across this sentry and asked what he was doing on the river bank? "Well Sir!" was the reply, "The Captain he said as I be to stop here and take care of this 'ere post and if any ship (sheep) came anigh 'ere, I be to fire on 'un!" Another story of Berkshire innocence I have heard, was of a recruit who had been "watching a friend drinking!" when required to take the oath swore faithfully to maintain "two religions" instead of "true allegiance" and nothing could persuade him to the contrary. Moreover being in a festive mood he would sing, not say, the words of the oath and was sworn accordingly.

The officers who went to Corfu were : Lieut.-Colonel Charles Bacon, Major Adam Blandy, Captains Sir Claudius Stephen Paul Hunter, Bart., William Richard Mortimer Thoyts, Charles Samuel Slocock, Richard Francis Bowles, James Douglas Christopher Deake Brickmann; Lieutenants Edward Tew Thomson, Henry Hanmer Leycester, Richard William Shackel, George William Bacon, George William Barker and Francis Reynell Cox ; Ensigns : William Stares, Charles Stuart Voules, James William Smith, Robert Tebbott, Henry Bayntun, George William Morland, and Henry James Lane; Surgeon : H. W. Reed ; Assistant-Surgeon : J. B. Alder ; Quarter-Master : John Milne.

The following officers remained at the Depôt at Reading : Captain Henry Greenway, Captain William Francis Wheble; Lieutenants : Arthur Deane and T. F. Maitland ; Ensign : Ward Soane Braham.

The services of the Regiment for Foreign Service were accepted in July but it was not until autumn that arrangements could be made for its conveyance thither. At last, in September they were sent by train to Portsmouth, and were finally embarked on the 22nd September, 1855.

The Report given by the Governor of Portsmouth was most gratifying and deserves to be recorded :—

" Major-General Breton, commanding the South-Western District, has ordered the commanding officer to express to the officers, non-commissioned officers, and soldiers of the Royal Berkshire Militia his great satisfaction at the appearance of the regiment, when inspected by him previous to embarkation. Their soldier-like appearance, steadiness, and the fact of there being no defaulters, prove to him that Her Majesty has selected a regiment worthy of Her Royal confidence, wherever they may be called upon to serve; and he should report to the General Commanding-in-Chief, for Her Majesty's information, that the Royal Berks Militia was the best regiment he had embarked for foreign service, and sincerely wished they would ever retain their justly-deserved commendation.

The Head-Quarters of the Regiment sailed from Portsmouth in the *Saldanha* on 22nd September, 1855, arriving on the 17th October following. Two companies of the Regiment, under Captain Thoyts, sailed in the *Great Tasmania* .in which was also conveyed the 3rd Middlesex Militia, under Lieut.-Colonel Glossop. During the voyage there occurred a single case of small pox on board, the victim being Sergeant Read, but he recovered before they reached Corfu and is now Sergeant of Police under his former commanding officer, Colonel Blandy.

In garrison at Corfu, were also the Oxfordshire and Wiltshire Militias and some Royal Artillery. Garrison duty was varied with shooting expeditions to the Coast of Albania, for woodcock, snipe, wild fowl, and boar.

The ladies of the Berkshire Militia who accompanied the Regiment were : Lady Hunter, Mrs. Bacon, Mrs. Cox, Mrs. Douglas and Mrs. Alder.

Lady Hunter gave a ball on the last night of the year 1855. The Fleet came to Corfu in January, 1856, consisting of the *Modeste*, the *Triton*, the *Sidon*, and *St. Jean d'Acre*, then followed more entertainments with a ball given· by Lord Methuen on January 16th.

Two captains, Slocock and Bowles, had obtained leave of absence and went to the Crimea, from whence they returned on January 18th, and a few days later Major Blandy went to Zante in command of the island. A large Fancy Ball was held at the palace on the 24th.

Then an Austrian man-of-war came into port and the officers were invited to dinner. Tableaux at the Palace seem to have ended the gaieties, at which there had also been a Fancy Ball and many smaller dances, given by Sir John Young, the Lord High Commissioner. On May 16th, the *Imperador* arrived, having on board the " Buffs " from the Crimea, and a few days later came the *Ripon* with the 68th Regiment. The Queen's Birthday was loyally celebrated and a *feu de joie* fired, the next day was Sunday. On Monday, May 26th, the Berkshire Militia bade farewell to Corfu ; at 5 o'clock that evening the *Imperador* sailed for England, being crowded with soldiers, for there were also 150 of the 3rd Middlesex men on board, besides the Berkshire Militia. The weather was stormy, few of the ladies appeared during the voyage. Off Algiers a fair wind began, to blow and Gibraltar was reached at 7 p.m. on June 2nd, where they had to put in to coal, after that it was wet and squally. A collision with a brig happened, luckily without damage to anything or anybody. Off Lisbon, on the 5th, there was sunshine but a rough sea, and all through the Bay of Biscay it was very rough, so much so, that there was no Sunday service. At last England once more appeared in sight, and the *Imperador* arrived at Spithead on June 11th at 3 o'clock in the afternoon. The following day, early, crowds assembled to see the first Transport enter Portsmouth Harbour. The band played merrily as they drew to the landing stage. By June 13th, the Berkshire Militia was back again at Reading in their old quarters in billets in the town. The officers being at the Great Western Hotel.

There had been a dark side to the picture, the time spent in Corfu had not been all gaiety and brightness. An epidemic of Cholera had broken out, the regiment lost about fifty men, and nearly the same number of women and children, most of whom are buried in the Island of Vido. The kindly feeling existing in the regiment is well shown by the fact that during all this trying time there was no lack of hospital attendance, the numbers at each morning parade who volunteered their services as hospital orderlies much exceeding the number who could be profitably employed on that duty. This sad time is mentioned in *The Militiamen at Home and Abroad* (by "Emeritus"), which was written by Major Prower, of the Wiltshire Militia, and gives an interesting account of Corfu in 1855-56.

CORFU from VIDO

Drawn in 1855 by Capt W R M Thoyts

VIDO from FORT NEUF

Drawn in 1855 by Capt W R M Thoyts

1856. May 8th. The order for disembodiment was agreed upon in Parliament, and the thanks of the House of Commons was voted to those regiments who had volunteered for service. Further to mark appreciation the Commander-in-Chief signified the Queen's pleasure that the word " Mediterranean " should be granted to be inscribed on the colours of the following regiments :—Royal Berkshire, for service at Corfu ; East Kent, for service at Gibraltar ; 1st Royal Lancashire, for service at Corfu and Zante ; 3rd Royal Lancashire, for service at Gibraltar ; Northampton, for service at Gibraltar ; 3rd Royal Westminster, Middlesex, for service at Corfu ; Oxford, for service at Corfu ; 1st King's Own, Stafford, for service at Corfu and Cephalonia ; Royal Wiltshire, for service at Corfu ; 2nd West York, for service at Malta.

The Berkshire Militia was not immediately dispersed after their return home, they were again billeted in the town of Reading ; the only event being a theatrical performance given by the officers in the Theatre.

July 4th. The men were paid off and dismissed to their homes after a year and a half embodiment, seven months of which had been spent out of England.

1857. The Indian Mutiny broke out, troops had to be sent from England to quell it. To supply their place the Militia were again embodied on the 1st October, and the Berkshire men were sent to Aldershot for duty, there they remained in the North Camp, until the following spring, May 7th. On their departure from the Camp, Major-General Hon. A. A. Spencer expressed his high approbation of their good conduct in quarters, and of the zeal of both officers and men during the period they had been under his command.

The non-commissioned officers and men left behind at Aldershot were attached to the 4th Royal Lancashire Militia. Colonel Blagrave gave his soldiers great praise for their high state of discipline and orderly behaviour.

1858.  Recruiting was actively carried on from this time for some years. Sergeants were stationed in the various large towns and villages in the county for the purpose. The acting Adjutant, Surgeon, and sergeant assistant, went round to all the places to attest volunteers. Newbury, Maidenhead, Aldermaston, Wokingham, Abingdon, etc., are thus mentioned.

1859.  This year the regiment did not assemble until July, a curious time of year to choose in an agricultural county where the hay cutting is quickly followed by corn harvest, and both usually are in progress at midsummer; no reason is given for this alteration of training, nor do I find any previous example of it, except in the autumn of 1769, the end of April or first week in May having been the usual time for assembling.

Colonel Conway, of the Grenadier Guards, inspected the regiment on August 2nd, and both he and Colonel Bacon expressed satisfaction with it in every way.

1860.  The following year the inspection was held by Colonel Lord Frederick Paulet, of the Coldstream Guards. Medals had been presented during the training by Colonel Blagrave to Quarter-Master Milne and Sergeant-Major Frederick Staden for long service and good conduct.

The clothing, etc., was given in after the training at the Forbury, and was stored in the gaol.

August 15th. Her Majesty the Queen graciously approved of the distinctive regimental badge of the regiment, which had been worn for many years.

The officers' cap badges of 1852 were embroidered in colours on the caps, and all varied according to the taste and fancy of the manufacturers; there was also some difficulty in getting them worked, so in Corfu a badge was designed by Captain R. F. Bowles,* and made in silver at the cost of twelve shillings each, which continued to be worn until 1880, when, most reluctantly, it was given up,

---

* This badge will be found stamped on the outside of this book.

and the regiment was obliged to adopt the "Dragon" of the 1st Battalion. Even the soldiers objected: "Be we to wear they cats on our collars?" they asked regretfully.

When the crest was altered in 1880 I sketched roughly a parody of it and sent it. to one of the officers. Some time afterwards, to my amusement, another officer gave me a photograph of it, remarking that speculations were rife as to who was the author of it. I did not enlighten him.

1861.   This year was marked for changes in the regiment. Colonel Blagrave and Lieutenant-Colonel Charles Bacon both resigned, and the command was taken by Lieutenant-Colonel Blandy with Colonel Charles Bacon as Honorary Colonel.

Lord Norreys, who had entered as Lieutenant in 1858 and resigned shortly after, was appointed senior Major by the Lord-Lieutenant, Captain Thoyts obtaining his promotion as second Major. Captains Slocock and Wheble, both of whom had served since the reorganisation in 1852, now resigned.

The Lord-Lieutenant of the county, the Earl of Abingdon, personally inspected the regiment on May 14th, and was received at the Great Western Station by a guard of honour with the regimental colours (the colours consist of the Queen's and regimental colours), at 10.20 that morning, and the parade took place the same afternoon on the Forbury. Year by year after this the annual training took place regularly at Reading until 1867, after which the regiment was sent to Aldershot for several trainings. They were inspected every year before being dismissed; no event of special interest worth recording seems to have occurred. The regiment was well up to strength, efficient and orderly, and in every way kept up its character and traditions.

1872.   In this year the strength was increased from 777 to 800 men.

1873. Early in this year Major Thoyts resigned, having been away from the regiment, and seriously ill, for two years. The majority was given to Captain Van de Weyer.

I have not so far alluded to the regimental plate, so will do so now. It is said that the old regimental plate was divided among the officers at the disembodiment in 1817, but I have not been able to trace any of it. Of the plate now possessed by the regiment, this illustration shews the beautiful centre-piece and dessert dishes presented by the ladies of Berkshire in 1855. In front of it, looking almost as if it were part of the centre-piece, stands the great tankard given by Lord Abingdon (formerly Colonel Lord Norreys) when his horse, "Sir Bevys," won the Derby. The two cups, with covers, were the gifts of Major A. W. Hay; they are copies from an old Irish cup. The two pipe-lighters, in the form of the regiment badge, were presented by Captain Thornton. Among other gifts, were those of Capt. Nepean, Captain Van de Weyer and others, a very fine service of fish knives and forks from Mr. R. Cazenove, lamps from Mr. R. Hargreaves, etc., etc.

# CHAPTER XII.

TWENTY-FIVE YEARS IN THE REGIMENT.—1873-1897

### BY LIEUTENANT-COLONEL BOWLES.

AVING been asked by the author of this Regimental History, as the present commanding officer, to carry on, as well as I am able, the history of the regiment from 1873 to the present time, I will endeavour to set down, from recollection and the records, something of its more modern history; an easy task when compared with the labour and research necessary for the first time to rescue from oblivion, to compile and put together in a readable form the scattered fragments of our county regimental history, and, in some degree, of the old Militia force (the constitutional defensive force) of the country at large.

This history will be, I believe, of the very greatest interest, not only to those at present serving, but to those who have served in it during more recent years, and those who may serve in it hereafter, and also, I venture to think, to very many Berkshire men and women, many of whom have had relations serving, or whose forbears served at one time or another. I also welcome this work, as I believe it will bring home to some a knowledge, or a more complete knowledge, of their special county regiment's history, its past and present status, and its modern intimate connection with its distinguished 1st and 2nd Battalions of the active British Army, in which many soldiers, trained for a time in its ranks, have continued their service, and to which we send recruits almost daily, receiving in return a percentage of good men who have concluded their active and first class reserve service. Anything that will bring home to our county the services of the regiment in the past, and its continuing efficiency at the present time will, I am sure, work for good, in causing more knowledge of the regiment, and more interest in it to be shown, to the increase of that *esprit de corps*, which is so necessary to a regiment, and the marked existence of which I recognise at the present time, both among ourselves and, I believe, in the county at large.

I will further express a hope that the result may be—and I would address myself here especially to all employers of labour—that more and more young men may be encouraged to come to us, rather than discouraged, when the many benefits in the nature of training, strict discipline, habits of cleanliness and order, and physical training are more fully perceived. Our only "defect," I was almost rash enough to say, but I will substitute "difficulty," is that we cannot obtain and keep a sufficient number of eligible recruits to maintain the numbers on the establishment, *viz.*, 901 of all ranks, including officers and Staff, of whom 842 are attested Militiamen. As an example of this I may say, that according to the annual inspection returns for the year, 1896, we were some 220 short of our number, a state of things that would have been seriously animadverted upon were it not that we had passed on almost that identical number during the recruiting year to the active army. Now the source of this difficulty is, without the shadow of a doubt, in the employers of labour, and it is for that reason that I have expressed a hope that young men may be encouraged, and not discouraged, in joining. At present an employer of labour invariably says: "If you join the Militia I will not employ you," for though it is *contrary to law* to discharge a lad *because* he joins, yet nobody can make an employer engage a Militiaman. Now I will give three good reasons for encouraging recruiting.

Firstly. Patriotism, which should be strong enough to overcome some personal inconvenience.

Secondly. The fact that the Ballot is still the law of the land, and there is great danger in these somewhat critical times, and in the face of England's comparative extreme weakness, face to face with the enormous modern armies of other nations, that if the absolutely voluntary system fails, the nation may have to enforce it, when all would be liable to serve in the defensive army of England, anywhere in the United Kingdom, though not out of it, except by their own consent. Would not this be considerably more inconvenient than the temporary loss of the services of an employé?

Thirdly. The benefits of training and discipline that have been referred to above.

My preface has been a long one, but I will not apologise for it, for I believe it was my duty to say what I have.

To introduce my story I must again become egotistical, and refer to myself, though I promise not often to offend again. I was gazetted in 1873, at a time when all recommendations for a commission in the Militia were made by the Lord-Lieutenant; now the Lord-Lieutenant may recommend on a vacancy occurring, but if he does not do so within one month the Commanding Officer recommends to the Secretary of State for War. This power of recommending for first commissions is all that remains to the Lords-Lieutenant of all their old control over the Militia forces.

At the time I was appointed I was at Oxford, and joined after the training of 1874 had begun, at Aldershot, in the old Guards' enclosure, now the grounds of the General's House. I was the only recruit officer, and was put through my facings day by day by a corporal of Scotch Fusilier Guards, behind a clump of fir trees, until I was supposed to be capable of carrying the colours and taking my turn on the duty Roster. The training of that year was extended to six weeks. The

o

regiment was commanded by Lord Norreys (now Lord
Abingdon and our Honorary Colonel). The other officers
were :—

| Joined. | Retired. | | Remarks. |
|---|---|---|---|
| 1862. | 1886. | Major V. W. B. Van de Weyer. | Afterwards in command. |
| 1857. | 1888. | Major J. Blandy-Jenkins. | Afterwards in command. |
| 1855. | | Captain (Hon. Major) W. S. Braham. | Instructor of Musketry. Is dead. |
| 1864. | | Captain R. C. Arbuthnot. | Is dead. |
| 1871. | 1883. | Captain W. M. C. Pechell. | Resigned & joined 3rd Northumberland Fusiliers. |
| 1870. | 1879. | Captain W. H. Morland. | Is dead. |
| 1871. | 1894. | Captain G. B. Eyre. | Afterwards as Lt.-Col. Archer-Houblon in command. |
| 1873. | 1877. | Captain C. C. Oldfield. | Captain (half-pay) late 38th Foot. |
| 1873. | 1875. | Capt. C. E. H. Vincent. | Late Lieut. 23rd foot. |
| 1871. | 1881. | Lieutenant F. J. Eyston. | Is dead. |
| 1871. | 1879. | Lieutenant F. M. Atkins. | |
| 1872. | 1874. | Lieut. Mowbray Allfrey. | To 15th Hussars, 1874. |
| 1873. | 1875. | Lieutenant G. A. C. Reid. | To a Highland regiment. |
| 1873. | 1875. | Lieut. W. H. Hippisley. | To Scots Greys, 1875. |
| 1873. | 1880. | Lieutenant C. V. Blythe. | |
| 1873. | | Sub-Lieut. T. J. Bowles. | Now in command. |
| 1873. | 1873. | Sub-Lieutenant W. H. Grenfell. | Did not join. |

Captain A. T. Pratt Barlow (joined 1861) resigned just
before the training; also Lieutenant S. A. Hankey (joined
1871), Lieutenant R. G. C. Mowbray (gazetted 1872, but
never joined), and Lieutenant Hon. C. C. Bertie (to the 47th

Foot, November, 1873). The Staff officers at this time were: Instructor of Musketry, Captain (Hon. Major) W. S. Braham; Adjutant, Captain Lang (late Captain 34th Foot); Quartermaster L. Milne (late 72nd Foot); Surgeon-Major, — Reed, M.D.

Major Blandy Jenkins had left the regiment as a subaltern soon after the Volunteer movement was started, in 1860, and served for a short time, as a subaltern, in the Abingdon Company of Volunteers, but soon returned to the Militia. Major Braham (who joined in 1855) was a brother of Lady Waldegrave, and one of the best mimics and most amusing of men; when in the mood he kept the whole mess in a roar. He lived by the river, close to Caversham Lock; and there returning, after a long absence, when the floods were out, to a damp house, he took a chill and died. He was for years Instructor of Musketry. Captain (now Sir Howard) Vincent joined from the 23rd Regiment as Captain; he was, at the time, reading for the Bar and studying French Law as well, and, after being called to the Bar, was for some time head of the Detective branch at Scotland Yard. After leaving us he went to a London Volunteer Regiment, the Central London Rifle Rangers, and has now for some years commanded the Queen's Westminster. He is M.P. for Sheffield, and married a grand-daughter of Mr. Morrison, of Basildon Park. Lieut. Hippisley is now second in command of the Scots Greys, and has been Adjutant of the Berkshire Yeomanry. Lieut. Grenfell, of Taplow Court, never joined.

There were manœuvres on a considerable scale at and around Aldershot this year,—following the very large manœuvres of the previous year on the Wylie, and about Salisbury Plain and Amesbury (at part of which I was present),—and the regiment took part in them. For the most part, these manœuvres were carried out from the Standing Camps, but the Royal Berkshire formed part of a force sent out to Woolmer Forest for two nights, from

O 2

whence it fought its way into Aldershot. The Royal Berkshire were brigaded with the 78th, 79th and 93rd Regiments,
the former commanded by Colonel Mackenzie, with Lieut.
J. Allin (who afterwards joined us as a Captain) as Acting
Adjutant. The 93rd Regiment had lately, come home after
a long tour of foreign service, and I well remember them
counter-marching by half-companies round a pivot, in the
old style, on a brigade parade. Major Wallingford Knollys
was Field Officer of the day when we reached Woolmer;
and, as Orderly Officer, I remember taking him into our
mess at night, after he had been his rounds, and finding
Major Braham entertaining General Herbert and his Staff,
who messed with us, with some story in which Major
Knollys figured, at the moment we lifted the curtain of the
tent. Major Braham got out of the difficulty by repeating
his story, which Major Knollys, of course, took in excellent
part. I remember another of Major Braham's jokes was
the following riddle: "When Generals Smith and Parke are
opposed to one another, to what well-known places in
London do they go?" The answer being: "One goes to
Hyde *Parke* and the other to Hammer-*Smith*."

The regiment had some hard work; and poor Walter
Morland, one of our Captains, during a day of outpost work
near the "Jolly Farmer," at Bagshot, who was posted on a
hill all day with no food, sustained a sun-stroke, which,
neglected at the time, caused his death a few years later.

Colours were carried on all occasions in those days, and
the subalterns who carried them had the best of good
times, generally lying down with the band in a wood, most
of the day. and sampling the contents of the luncheon
cart.

This was Captain Lang's last training as Adjutant; he
had served in that capacity under the old regulations for
many years, was rather fussy, but a smart officer. Some
young officers, I think of the Guards, played a trick on him

which, not unnaturally, made him very angry. He received what purported to be a Divisional order, that the Royal Berkshire were, on a certain day, to find a party "to carry coal for the Guards!" His indignation was immense, but of course the joke was not carried too far, and he allowed himself to be appeased. Another story of him was that, on one occasion, having forgotten something, he returned late at night to the old Head-Quarters in Mill Lane, but had forgotten the "Countersign;" and the sentry, coming down to the charge within an inch of the centre button of his great coat, refused to let him pass, and he never got in that night.

The Mess at this time, during the recruit training, was held at the "Queen's Hotel;" but, on Assembly, the regiment always went to Aldershot, partly to avoid the inconvenience of having the men in billets all over the town.

One incident of this training ought to be mentioned, viz.: On the return from Woolmer to Aldershot, the Highlanders leading (they were all well-seasoned men, not boys) and the Royal Berkshire being the rear battalion of the brigade, the distance, five miles, to a certain bridge near Frensham, which it was important to occupy, was accomplished by the leading battalion in fifty-five minutes, the Royal Berkshire not losing distance.

The Commanding Officer, having the option this year, preferred not to have chakos for the men, but an extra glengarry instead; the chakos having to last many trainings, and naturally, in tents and billets, often getting sadly out of shape. The Commander-in-Chief, however, remarked upon our want of chakos at a review he held the same year at Aldershot. No chakos or helmets have been supplied to the men since, except specially for the Jubilee Review in 1887, till this year, 1897.

The well-known Mutton, of Brighton, messed the officers this year.

The regiment was well reported on, except that it was 237 under strength.

1875.    The regiment trained again at Aldershot. Lieut. Reid was Acting Adjutant, in place of Captain Lang, and Lieutenants Allfrey and Hippisley had gone to the 15th Hussars and Scots Greys respectively. The new appointments were E. Vincent (now Sir Edgar), J. W. Allin (from 78th Foot to be Captain), and E. G. R. Hopkins.

The Assembly was on May 10th, and the regiment proceeded the same day to Aldershot. It lay on Cove Common, and was attached to the 3rd Brigade, under Major-General A. J. Herbert. The regiment took part in the Queen's Birthday Divisional Parade on May 29th; also on June 2nd, in a Divisional Parade of Militia, when it was inspected by the Inspector-General of Auxiliary Forces, and on the same day by Major-General Herbert, C.B. On June 4th, it took part in a Divisional Field Day, when His Royal Highness the Commander-in-Chief was present, who expressed his satisfaction at the appearance and steadiness of the regiment. It returned to Head-Quarters on June 5th and was dismissed the same day.

The recruits, previous to the training, were called up for fifty-six days' preliminary drill, and then accompanied the regiment to Aldershot. Very curiously the regiment was again exactly the same number (237) under the strength of the establishment as the previous year, though there were 136 recruits present with the regiment.

The Establishment at this date was 903 of all ranks. The variations in the Establishment at various dates, from 1859, may here be noticed: In 1859—869 of all ranks; 1860 —873; 1861-64—866; 1865-66—688; 1867—687; 1868— 866; 1869-71—867; 1872-3—890; 1874—904; 1875—903; 1876—902; and it now is, and has been for many years, 901.

During the years 1865-67 the strength of the regiment was reduced (by a letter from the Secretary of State for

War, dated August 23rd, 1864) till further orders, to eight
sergeants, twenty-four corporals, and 600 privates (Militiamen). The officers and Staff remaining practically the
same.

In pursuance of an order from the Secretary of State for
War, dated War Office, October 16th, 1867, the regiment
was directed to enroll recruits to the full strength of its
quota, to be completed during the next two years.

1876.  The regiment was ordered, on July 3rd, to take
part in large manœuvres, and formed part of a
Militia Division encamped on Minchinhampton Common,
a high plateau near Stroud in Gloucestershire, remarkable
for extensive traces of old earth dwellings. It belonged to
the 2nd Brigade, under Major-General J. E. Thackwell,
C.B. (who at one time rented Southcote Lodge, near Reading), forming part of the 3rd Division, under Prince
Edward of Saxe-Weimar, 5th Army Corps for mobilisation,
Army Corps' Head-Quarters were at Salisbury. Though
nominally mobilised there was no transport, and the
Division remained in a Standing Camp, but the tentage was
allotted on a manœuvre footing, and the baggage was
restricted. The weather was generally extremely hot, and on
the occasion of a Review by the Commander-in-Chief
numerous slight cases of sunstroke occurred, men falling
in the ranks as he rode down the lines, whereupon His
Royal Highness at once dismissed the parade. There were
magnificent views from the Common but absolutely no
shade.

Our beef, which consisted of Spanish cattle brought from
Corunna, and which ran on the Common, was of the
toughest, something like a " trek ox " at the Cape, as it had
to be killed and eaten the same day.

In the Division, beside ourselves, were the two Gloucester
Battalions, the Oxfordshire, the Wilts, and the two Tower
Hamlet Battalions, then red regiments. Old feuds between

Line regiments are not uncommon, but it is singular that the same thing should occur between Militia regiments who may not have met for years; however, something of the kind occurred here, and came to a head one Sunday, between some of our neighbours on either side, probably because the men had nothing to do; but by the exertions of their own officers and by a double line of sentries being run across our Camp between them by the two Majors and the Adjutant, who were in Camp, any serious mischief was prevented. Officers and strong pickets were posted at night down in the surrounding country and near the entrance to the town of Stroud.

Some large divisional sports were organised and successfully carried out. Officers' sketching parties were sent out round the Camp, though, I am afraid, many of us were hardly as well up in that part of our duties as we might have been. This was Captain Murphy's first training as Adjutant; he was the first of the new order of Adjutants appointed to us from the now 2nd Battalion, the old 66th Regiment.

The regiment was inspected by Major-General H.S.H. Prince Edward of Saxe-Weimar, C.B., who expressed his great satisfaction with its efficient state.

On the last day of the training, July 29th, the regiment marched at 4.30 a.m., entrained at Stonehouse Station, and proceeded to Reading, where clothing and arms were taken in and the men marched to the trains for their various destinations on the same day. I believe we were the only regiment who at that time adopted this plan of sending the men home, which has since been ordered to be done generally. It certainly had the effect of passing the majority of the men to their homes with their money in their pockets instead of their being tempted to spend it in the town, very often to the disadvantage of good order and discipline.

Since the training of 1875, Captain C. E. H. Vincent had resigned, and also Lieutenant G. A. C. Reid, Lieutenant

Eyston was promoted Captain, and A. W. Craven, C. R. Hunter never joined, C. L. M. Pearson, and G. F. Clayton East were appointed. Captain T. Murphy, 66th Foot, was appointed, September 20th, 1875, as Adjutant.

Captain Murphy rejoined the 2nd Battalion Berkshire Regiment in September, 1880, for a short time; after which he retired with the rank of Lieut.-Colonel. He resided until lately in Reading.

1877. This year is to be remarked as being the first under the Brigade or Regimental District system. At this time the new Depôt Barracks in the Oxford Road, not then in the Borough of Reading, but in the Parish of Tilehurst, Berkshire, were called the "Depôt Centre;" they were afterwards called the 41st Brigade Depôt; and finally the District was called the 49th Regimental District, its Head-Quarters being at the new Barracks, where are also the Depôt of the 1st and 2nd Battalions Royal Berkshire Regiment and the Head-Quarters of the 3rd Battalion; there also are the whole of the arms, stores, and clothing for the Depôt, the 3rd Battalion, and for some of the Reserve.

Formerly the County had to find Head-Quarters and stores for the Militia. The following in this place may be of interest, and it may be premised that it is only of late years that Reading has become "The County Town."

Berkshire to wit. { At the General Quarter Sessions of the Peace, of our Sovereign Lady the Queen, holden at Abingdon, in and for the said County, on Monday, the Sixteenth day of October, in the Seventh year of the Reign of our Sovereign Lady Victoria, etc., etc., and in the year of our Lord 1843.

Before: Robert Palmer, Esquire, Chairman; Percy Henry Crutchley, Charles Eyston, Esquires; and others, etc., etc.

Messrs. Scott and Moffatt having obtained the sanction of the Inspectors of Prisons to the proposed Plan for

providing a Depôt for Militia Stores in the basement of the
new Prison, and a Guardroom, etc., on the outside of the
new boundary wall, with a covered way to connect the
Guardroom with the Depôt; it is resolved by this Court,
that the said Plan be adopted, and that the necessary works
be forthwith completed; and the Clerk of the Peace is
directed to transmit a copy of this resolution to the
Adjutant of the Berkshire Regiment of Militia, and to
inform him that possession of the new buildings will be
given to him with the least possible delay. By the Court,

*(Signed)* GEORGE B. MORLAND.

*Clerk of the Peace.*

At this time, 1877, the Militia Depot and stores were, and
had been for a long time, in Mill Lane, Reading, and so
continued till after this year, as the Barracks were not ready.
The recruits however, on assembling, to the number of 178,
on April 30th, were encamped in the field in rear of the Barracks, where the regiment is now-a-days usually encamped
and drilled, in order to get the men out of billets. The recruits
were inspected by Colonel T. Maunsell, C.B., commanding
41st Sub-District, on June 22nd, and proceeded with the
regiment on June 25th, the day of assembly, to Aldershot.
The regiment was attached to the 2nd Brigade, 3rd Division,
Army Corps, commanded by Colonel G. F. C. Bray. It was
encamped on Rushmoor, and lay next to the 49th Regiment,
to which regiment were attached the Depôt Companies 66th
Regiment, henceforth the 1st and 2nd Battalion Berks
Regiment. The regiment took part in some of the
operations during the manœuvres in the neighbourhood of
Aldershot, chiefly on the Fox Hills, and was inspected by
Colonel Maunsell, C.B., on July 18th. It returned to
Reading and was dismissed on July 21st.

There was some heavy rain during the training, and the
camp being on a hillside the tents were often flooded

one officer finding most of his portable kit had floated out of his tent during the worst night! The 49th and 66th Regiments were most hospitable. Colonel Fitz-Gerald commanded the former at that time. Otway, (afterwards for many years in the 3rd Battalion), Holden, (afterwards our Adjutant), Haggard, (now Major, who put us through musketry that year), and many others were with them. Otway at the time possessed a loud ticking American clock, which so disturbed his rest, that one night he hurled it far into the neighbouring plantation, where Holden, next morning, found it ticking as loudly as ever, annexed it, and had it for many years afterwards. He used to say nothing could stop it.

The rest of those officers who were not for early parade was much broken by the voice of the Drill Sergeant of the 49th Regiment, just behind their tents, drilling recruits. The owner of the voice, Sergeant Owens, one of a large family of soldiers, was afterwards on the 3rd Battalion Staff, and then in the Berkshire Police.

Since the last training Captain and Hon. Major W. S. Braham had died, and Captain Oldfield had resigned. A. Hargreaves, L. S. B. Tristram, T. J. Wheble, C. H. Ashurst and E. Davis were appointed. Lieutenant E. G. R. Hopkins resigned on July 18th this year.

1878. The recruits (254), with a proportionate number of non-commissioned officers and trained men for duty, assembled on March 18th, and, with the exception of H Company, who were billeted outside, were quartered in the new Barracks. They were inspected by Colonel T. Maunsell, C.B., Commanding the 41st Sub-District, on May 10th.

The regiment assembled on May 13th, and was placed under canvas in the drill field. This was the first training at the new Depôt and the first carried out at Reading for some years. The musketry was carried out on Coley Range.

The officers' Mess hut was not then built, and the officers tents and Mess were up in the Camp, taking off from the available space for drill.

The regiment was then, and until 1881, officially known as the Royal Berkshire Militia.

Since the last training, in addition to Lieutenant Hopkins, Quarter-Master J. Milne had retired, with rank of Captain on April 1st, having served as Quarter-Master since 1855. He came from the 72nd Foot. Quarter-Master-Sergeant C. Lewis was appointed Quarter-Master on the same date; he joined on March 3rd, 1868, from the 56th Foot. Lieutenant E. Vincent had resigned to the Coldstream Guards; he is now Director of the Ottoman Bank. C. Ashurst to Carnarvon Rifles, thence to 35th Foot. The Hon. M. C. F. Bertie (now Lord Norreys), A. Knox, C. W. Darby-Griffith, and N. H. Vansittart had been appointed.

1879. This year the recruits assembled on March 17th B and D Companies were in billets, the remainder in Barracks. They numbered 240 and were inspected by Colonel J. Jordan, C.B., on May 8th. The regiment assembled on May 12th and was placed under canvas in the drill field. It was inspected by Colonel Jordan, C.B., Commanding 41st Sub-District on May 29th, and was dismissed on the 31st. The regiment was favourably reported on.

Captain Blyth had, at this time, taken up road coaching and had his horses at Reading, making up his teams preparatory to running the coach from Oxford to London. He usually met the train leaving Paddington at 9 a.m., and brought up any officers who had had leave for the night to London. His coach was on the road before the end of the training; and the first day on the road all the officers who could get away, went to Oxford, and Lord Norreys drove the coach to London. The officers afterwards dined together. This was the first beginning of the Militia

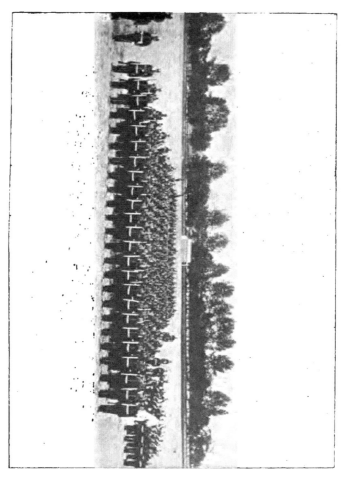

ON PARADE.

Regimental Dinner, which, by invitation of the 1st and 2nd Battalions, was afterwards amalgamated with theirs and one dinner club formed for the three battalions. This dinner takes place in Derby week.

Since the last training, Captain Morland (who was on leave in 1873), Lieutenants C. L. M. Pearson (to Rifle Brigade), G. F. Clayton East (to 25th Foot), L. S. B. Tristram (to Welsh Regiment), had resigned, and Lieut. A. Hargreaves had died; while K. P. Burne, K. Apthorp, C. S. F. G. Toogood and F. V. Allfrey had been appointed.

1880. The recruits (227) assembled on March 15th; G and H Companies in billets, the remainder in Barracks. They were inspected by Colonel Jordan, C.B., on May 5th, and went with the regiment to Aldershot on May 10th. The regiment assembled on May 10th and proceeded to Aldershot, where they were encamped on Redan Hill and attached to the 2nd Brigade, commanded by Major-General W. G. Cameron, C.B. The regiment was exercised, after a course of company and battalion drill, in brigade, and took part in the Queen's Birthday Parade.

Major-General Cameron gave a lecture in the Mess to the officers and examined them *vivâ voce*. He inspected the regiment on Rushmoor on June 3rd, and expressed himself well satisfied.

The regiment was this year stronger in numbers than it had been for years, or than it has ever been since. In fact recruiting was stopped before the training, as no supernumeraries were allowed to be enlisted, with the result that at the training the regiment was not quite up to strength. I had ninety-six men present with my company. At the Birthday Parade the regiment was in a second line in rear of the 52nd Regiment (at that time very weak, having lately come home). I believe they only mustered 132 files, as strong as possible, so that we overlapped them tremendously.

In marching past in column of double companies I
remember the lookers on gave us great *kudos,* but
accused us of copying the old Highlanders trick, by
holding the skirt of the next man's frock between the
fingers, instead of the kilt.   The regiment returned to
Reading and was dismissed on June 5th.

Since last training Captains C. V. Blythe and F. M.
Atkins, Lieutenants A. W. Craven, E. Davis (to 81st Foot),
and C. W. Darby Griffiths (to Grenadier Guards) had
resigned, and E. G. Costobadie, J. G. R. Homfray, A.
W. H. Hay, and F. M. Birch had been appointed.

1881.     The recruits (224) assembled on March 21st.
The B, D, G, and H companies were in Barracks,
and the A, C, E, and F companies in billets and lodgings
outside.   Captain Murphy had rejoined the 66th Regiment,
and Captain C. P. Temple had been appointed Adjutant
from the 49th Regiment in September, 1880.   The recruits
were inspected by Colonel Jordan, C.B., Commanding 41st
Regimental Sub-District, on May 11th.   The regiment
assembled on May 16th, and was placed under canvas in
the Drill Field.   Colonel Jordan made his inspection on
June 9th, and the regiment was dismissed on June 11th.
Lord Norreys had resigned on October 26th, 1880, and
Major Van de Weyer succeeded him as Lieutenant-Colonel.

The year 1881 is especially to be remembered, for it
was then the Territorial System came fully into
operation, and the regiment lost its title of "Royal," and
became 3rd Battalion (Princess Charlotte of Wales')
Berkshire Regiment, by general orders, on June 30th, 1881.

Extract of Horse Guards' letter $\frac{6109}{1557}$

"Horse Guards, November 28th, 1881.

"A stag under a tree has been sanctioned as the badge for the
helmet plate and glengarry of the Berkshire Regiment."

The following, though of later date, should follow here.

Regimental Orders: " Reading, January 9th, 1882.

"The following extract from Regimental District Orders is just published. Paragraph 1, dated January 9th, 1880. ' It has been decided by H.R.H. The Field Marshal, Commander-in-Chief, that the "China Dragon," as worn by the 1st Battalion of the Berkshire Regiment (late 49th Regiment), is to be the badge for forage cap of officers of the Berkshire Territorial Regiment.' "

All this actually came into force after the training, but in the Inspection Report for this year the regiment is, for the first time, called the 3rd Battalion Princess Charlotte of Wales Berkshire Regiment.   It will be noticed that the device of the Royal Berkshire Militia, " The stag under a bare oak," was adopted for the helmet plate and glengarry for the territorial regiment, and the " China Dragon " of the 49th Regiment for the forage caps.   The 3rd Battalion now changed its facings to white (all English regiments, not Royal, wear white facings), with a red mess waistcoat, until, as is recorded later on, the Royal title was restored for the services of the 1st Battalion in Egypt, where the Berkshire lads proved themselves excellent soldiers.   Gold lace was substituted for silver on the officers' uniforms, which was to be exactly the same as the Line Battalions, with the addition of M on the shoulder strap.   Silver lace had been made the regulation for Militia in 1830, and before that both Line and Militia wore gold and silver indiscriminately.

This year, the Officers' Mess hut was built inside the Barrack Square, thanks in great part to the generosity of Colonel Van de Weyer, and used for the first time ; it has since been added to and improved.

Since the last training, in addition to Lord Norreys and Captain Murphy, Captain Hon. Major Arbuthnot and Capt. Eyston, Lieutenants K. P. Burne (to 72nd Highlanders) and K. Apthorp (to 18th Foot) had resigned; and Major Lord Algernon M. A. Percy, Captains Temple, T. W. R.

Adams and J. T. F. Otway, and Messrs. F. W. Leybourne
Popham did not join, G. W. Thomas Viscount Savernake
and M. H. Burne were appointed. It will be seen there
were a good many changes. Lord Algernon Percy, who
had been Adjutant of his battalion of the Grenadier Guards,
came in as Junior Major. Captain Adams had been in the
Surrey Militia, but was living in Berkshire at this time.
Captain Otway had been in the 5th Dragoon Guards, and
afterwards in the 49th Regiment. Captain Eyston was
the only son of Mr. George Eyston, of Stanford Place,
Faringdon, and nephew of Mr. Eyston, of Hendred; he is
dead.    Captain and Hon. Major Arbuthnot had been in
the regiment since 1864.  His sight failed him; he is also
dead.

The officers this year gave an afternoon dance in the
new hut.

H.R.H. the Commander-in-Chief, in his observations on
the Inspection Report, says: " The report reflects credit
on all concerned."

1882.    The regiment assembled for its annual training
on July 24th and proceeded the same day to
Aldershot.  It was this year armed with the Martini-Henri
Rifle, in place of the Snider.

The old recruit training was this year done away with,
the recruits henceforward being drilled with the Line
recruits at the Depôt, on attestation; and, on completion
of the recruit course, dismissed to their homes, to come up
again with the regiment, with the exception of those enlisting shortly before the training, and whose recruits' course
would not be finished on the assembly of the regiment.
These men do an extended course at the Depôt, and do not
join the regiment till the succeeding training.  The Staff
were to be utilised for all purposes at the Depôt.  No doubt
this change has operated in giving more recruits to the Line
from the Militia Battalions.

The regiment, at Aldershot, was attached to Major-General Spurgeon's Brigade and encamped at Rushmoor. Major-General Spurgeon made his inspection on August 16th. The regiment returned to Reading on the 19th and the men were dismissed to their homes the same day.

It is believed that the Royal Berkshire was the first Militia regiment to proceed on the day of assembly to their destination, when training away from Head-Quarters, and to dismiss on the day of return. This was inaugurated during Colonel Blandy's period of command.

Lieut.-Colonel Van de Weyer was absent on sick leave from the training, and Major Blandy-Jenkins fell ill during the training and had to be moved to the "Queen's Hotel," the command devolving on Major Lord A. Percy.

While the regiment was at Aldershot the 19th Hussars, who had been made up to War strength by drafts of men and horses from other regiments, and who were encamped close to us on Rushmoor, marched out to embark for Egypt.

Some scratch sports for the men were carried out one afternoon. When at Reading they are an established institution.

The following officers had resigned since the last training: Captain Allin, Lieutenants A. Knox, J. G. R. Homfray (to 1st Life Guards), F. W. Leybourne Popham (he had never joined); and Messrs. J. C. Blagrave, F. Pratt Barlow and J. Blandy-Jenkins had been appointed.

1883. Training under canvas in the drill field at the Depôt; Assembly, May 7th. Inspection by Colonel the Hon. W. H. Herbert, commanding 49th Regimental District, on May 31st. The men were dismissed to their homes on June 2nd.

The regiment was well reported on; but there were an unusual number of absentees, and the regiment was 178 below strength.

P

Since the training of 1882, Lieutenant Toogood's services had been dispensed with.   Lieutenant E. G. Costobadie had resigned (to the Royal North Lancashire Regiment); and Captain John Preston, who had served on the Gold Coast and at the Cape (medal), had been appointed, with Messrs. F. Theobald, A. G. Egerton and T. F. J. L. Hercy.

1884.      As there was a good deal of small-pox in and around Reading this year, it was decided by the authorities that the training should not take place.   The following letter was received :

" Horse Guards,
" War Office, April 22nd, 1884.

" Sir,—With reference to your minute of 7th instant, and previous correspondence, I am now directed by the Field Marshal Commanding-in-Chief to inform you that, under the circumstances represented by the officer commanding the 3rd Battalion Berkshire Regiment, and as small-pox is still prevalent in the neighbourhood of Reading, the Secretary of State for War, in consultation with His Royal Highness, has decided that the above mentioned battalion shall not be called out for training in the present year, and I am to request that you will have the goodness to issue instructions accordingly.   Steps will in due course be taken to obtain an order by Her Majesty in council dispensing with the usual training in the case of the battalion referred to.

*(Signed)* JOHN ELKINGTON,
*D.A. Gen.*

Captain W. M. C. Pechell had resigned on January 8th, 1883, he was afterwards appointed to the 3rd Northumberland Fusiliers.   During 1884, Quarter-Master C. Lewis had retired (January 23rd) with the rank of Captain. Lieutenants F. V. Allfrey, M. H. Burne (to Royal Sussex Regiment), F. Theobald (to Berkshire Regiment) and A. G. Egerton, resigned.  Lord G. M. Pratt (late Grenadier Guards and West Kent Yeomanry) was appointed Captain ; Messrs. C. C. Leveson-Gower, A. H. W. King, J. R.

Gray, H. R. Homfray, and F. Porter were appointed, the last-named being transferred during the year to the 3rd Battalion Bedfordshire Regiment. Lieutenant and Quarter-Master Hollyer was appointed from the 1st Battalion.

1885. In this year the regiment resumed its old title of Royal, the Berkshire Regiment having been granted the title of "Royal" for the services of the 1st Battalion in Egypt. The training took place at Reading, in camp on the Drill Field, from May 18th to June 13th. Colonel A. Brown, commanding 49th Regimental District, inspected the regiment on June 11th. Lieutenant F. B. Pratt Barlow resigned to the Dorsetshire Regiment May 21st. The following were appointed before the training: F. T. Stewart, C. C. Williams, and D. Blake Maurice. This was Major Temple's last training as Adjutant, than whom no regiment of Militia ever had a better Adjutant. Major Temple rejoined the 1st Battalion July 30th, 1895, served with · it in Egypt, succeeded to the command, and commanded it also at Cyprus and Malta, where the 1st Battalion, keeping up the high character it had earned in Egypt, was remarkable for its excellent conduct, discipline, and smartness. It was composed almost entirely of Berkshire men, a large number of whom had passed through the 3rd Battalion, as I found when I visited Malta, on leave, in 1892. Colonel Dickson had then succeeded Colonel Temple in command. The battalion was 1050 strong, and about 80 per cent were Berkshire men, some of those who served with the 1st Battalion in Egypt and the Mediterranean have since been in the 3rd Battalion. This, in my opinion, is as it should be. Colonel Temple had a great opinion of the soldier-like qualities of the Berkshire men; he has since commanded the Regimental District at Worcester, and is now Assistant Adjutant-General at Cork.

1886. The regiment assembled for training at Reading on May 10th, and was placed under canvas in the

Drill Field. It was inspected by Colonel A. Brown, commanding 49th Regimental District, on June 3rd, and the men were dismissed to their homes on June 5th. Lieutenant-Colonel Van de Weyer, having resigned on April 4th with a step of honorary rank, Major J. Blandy-Jenkins was promoted to the command. Colonel Van de Weyer had served in the regiment very nearly twenty-four years. Captain G. B. Eyre was appointed Major. Captain Hon. M. C. F. Bertie (now Lord Norreys) resigned April 9th. Lieutenant George W. Thomas Viscount Savernake resigned April 11th, 1896, C. C. Leveson Gower (to Royal Warwickshire Regiment) April 29th, J. R. Gray (to Royal Irish Rifles) November 24th, 1885, and H. R. Homfray (to Royal Irish Rifles) same date. Captain H. Winton Holden was appointed adjutant from the 1st Battalion on October 1st, 1885, and Messrs. John E. Duffield and H. J. M. Cleminson were appointed.

1887. The regiment assembled for training at Reading on June 20th and proceeded the same day to Aldershot, where it was attached to the 3rd Infantry Brigade under Major-General Buchanan, C.B., and encamped on Cove Plateau. Major Lord Algeron Percy having been transferred on June 24th, 1886, to the 3rd Northumberland Fusiliers, I was appointed Major. The regiment formed part of the 8th Brigade, 4th Division, 2nd Army Corps. On July 9th, on the occasion of the Queen's Jubilee Review in the Long Valley, the 2nd Army Corps was commanded by Major-General Sir Evelyn Wood, V.C., G.C.M.C. At the conclusion of the review it formed a part of the force which lined the road on each side by which the Queen returned to the Royal Pavilion. The regiment was inspected on July 12th by Major-General Buchanan, C.B., and returned to head-quarters on July 16th and was dismissed. The Inspecting Officer made a very thorough inspection of the regiment this year and reported very favourably; he was not, however, satisfied with the musketry system, and noticed the worn out condition of the great coats, he also recommended that helmets should be issued. The great coats were soon afterwards gradually replaced, they have since been much better. Helmets were served out specially for the Jubilee review and then returned (there was little time to fit them). During 1896 they were issued to the regiment and are available for this (1897) training. Our old comrades, the 4th Oxfordshire Militia, were encamped next to us on Cove Plateau; the two regiments held sports together. The Berkshire Yeomanry took part in the Jubilee review and made a halt at our camp on their return, to refresh men and horses, but had to leave before the regiment returned. The men of the regiment had extra delicacies provided for them on the Queen's Jubilee day.

In addition to Major Lord A. Percy, Lieutenants T. F. J. L. Hercy, F. Stewart (to Highland Light Infantry), and H. J. M. Cleminson had resigned since the last training, and Messrs. Sidney T. Hankey, F. H. G. Hercy, Baldwin Hodge (resigned before the training), C. E. M. Y. Nepean, F. G. Barker, Arthur Jocelyn Charles Viscount Sudley, J. E. Rhodes and W. Thornton had been appointed.

139 men volunteered for the line.

1888.     The regiment assembled for training at Reading on May 14th, and was placed under canvas in the Drill Field. It was inspected by Colonel A. S. M. Browne, commanding 49th Regimental District, on June 7th in Review Order; on June 8th in Marching Order. He expressed himself well-pleased with what he saw, but thought care was required in fitting the new valise equipment; hitherto the old pack or knapsack had been carried. The men were dismissed to their homes on June 9th.

Colonel Blandy-Jenkins, who had been granted a step of honorary rank January 29th, 1887, resigned immediately after the training. Major Eyre (who afterwards assumed the name of Archer-Houblon) was appointed Lieut.-Colonel on the same date, June 16th. Captain Preston had resigned since the last training (he was afterwards appointed a magistrate in Ireland), and also Lieutenants J. Blandy-Jenkins and C. C. Williams. The following had been appointed: E. M. Fowler, G. C. T. Willes, and P. L. Leveson Gower.

The Battalion furnished 128 men to the Line.

1889. The regiment, as last year, was placed under canvas in the Drill Field at the Depôt, assembling on May 6th and being dismissed on June 1st. It was inspected on May 30th and 31st by Colonel Browne, commanding the 49th Regimental District. The regiment was well reported on and, in addition to furnishing 145 men for the Line, was stronger in numbers by eighty-nine men.

This was Lieut.-Colonel Eyre's first training in command; Captain T. J. Wheble had been appointed to the vacant Majority; Captain King had resigned, February 19th, 1889; Lieutenant F. H. G. Hercy had been transferred to the 3rd Royal West Surrey Regiment; J. R. Gray returned to the Regiment as Captain; and G. E. Phillips had been appointed.

1890. The recruits assembled on April 28th for a course of instruction in musketry before the training; this was a new departure this year. The training again took place at the Depôt. The regiment assembled on May 5th, and was dismissed on May 31st. It was inspected by Colonel H. C. Borrett, commanding 49th Regimental District, on May 29th in Review Order, and on May 30th in Marching Order. Colonel Borrett expressed himself as much pleased with all he saw.

Lieutenants J. E. Duffield, Arthur J. C. Viscount Sudley (to Royal Horse Guards), J. E. Rhodes (to King's Royal Rifle Corps), and G. E. Phillips had resigned; and Percy

Downes, W. J. B. Van de Weyer, Hon. R. C. Craven, E. T. Whitehurst, Robert Hargreaves, and W. A. F. White, had been appointed.

Lieutenant Sidney Thornhill Hankey was appointed to 2nd Life Guards directly after the training.

Second Lieutenant E. T. Whitehurst was drowned at Shrewsbury on July 18th while gallantly endeavouring to save life, a very promising young life prematurely cut short.

1891.   Major H. W. Holden had joined the 2nd Battalion on October 30th, 1890, on the completion of his period of service as Adjutant, and Captain C. Mackenzie Edwards had been appointed from the 1st Battalion.

The regiment assembled for training on May 11th and proceeded the same day to Aldershot, where it was encamped on Watts's Common, being attached to the 1st Infantry Brigade, under Major-General Gregorie, C.B., by whom it was inspected on June the 4th and 5th, in Drill, the Attack Formation, and Interior Economy. The regiment returned to Head-Quarters on June 6th and was dismissed.

The work of the training, after the first three days of great heat, was interfered with by continuous cold and wet weather. The musketry was carried out on the old Cæsar's Camp ranges, where one man was hit by a party of a Cavalry Regiment firing on the adjoining range; he only received a flesh wound.

The Inspecting Officer expressed himself well satisfied with the progress made in spite of the wet weather. 141 men volunteered for the Line.

Captain, Hon. Major J. W. R. Adams was transferred to the 4th Middlesex Regiment, September 19th, 1890; Lieutenants D. B. Maurice to the 1st Battalion, and E. M. Fowler had resigned; and S. H. Rickman, R. F. Cazenove, M. M. Mercer Adam, and C. Knipe had been appointed before the training, while S. L. Barry was appointed directly afterwards.

1892.      The recruits were again assembled for fourteen days' preliminary musketry before the training. The regiment assembled at Reading on May 9th, and was dismissed on June 4th.

During the training, Major General Lord Methuen, commanding the Home District, inspected the Depôt, he watched the 3rd Battalion drill and march past, and expressed himself as extremely well satisfied with what he saw.

Colonel H. Borrett, commanding 49th Regimental District, inspected the regiment at Battalion Drill and went through the Books, he also saw the regiment in extended order on a Field Day at Englefield Park, which Mr. Benyon had most kindly lent for the occasion. While on the return march through Theale, a short but very heavy thunderstorm gave the regiment a good drenching. With a view to not disturbing the deer and game, no blank ammunition was used. The Interior Economy of the regiment was well reported on.

Lieutenant P. Leveson Gower resigned (to Sherwood Foresters), September 9th, 1891, and H. V. Rhodes was appointed. The numerical strength of the regiment had somewhat increased.

1893.      This year the regiment was ordered to get through its musketry course and then, as one of the regiments of the 18th Brigade (2nd Army Corps) in the Home Defence Scheme, to proceed to Forest Row, Ashdown Forest, for brigade drill and manœuvres. Accordingly the recruits assembled for preliminary musketry on April 15th; and the regiment on May 1st. After the musketry course, carried out for the last time on the Coley Range with the Martini Rifle on May 16th, the regiment entrained at Reading and was conveyed to Forest Row station. It was encamped, together with the 3rd Royal Sussex, the 3rd Royal West Surrey and the West Kent, about one-and-a-half miles from the station, on Ashdown Forest, under Brigadier-General Fowler Butler, commanding 18th Infantry Brigade.

The regiment was employed with its own brigade and brigades of Regular troops in various Field Days and manœuvres, and in outpost work both by day and night. The regiment was inspected on parade by Brigadier-General Fowler Butler, and the books, etc., by Colonel Borrett, commanding 49th Regimental District.

H.R.H. the Commander-in-Chief witnessed the concluding day's operations and a march-past. The regiment was very well reported on, it returned to Reading on May 26th, where the camp had been left standing. The men were dismissed to their homes on the following day. This was Colonel Archer Houblon's last training.

There were no resignations since last training and no new appointments.

1894. A short time previous to the training, Colonel Archer Houblon resigned the command, and I was appointed Lieut.-Colonel, to hold the appointment for five years. This was the first appointment under the new Time Warrant.

The regiment was ordered to train at Reading in May and June. The recruits' musketry course and that of the battalion to be conducted at Churn, which Lord Wantage had placed at the disposal of the authorities for that purpose.

The regiment was this year armed for the first time with the Magazine Rifle, and for that rifle the range at Reading was pronounced unsafe to the public. A new valise equipment was also issued.

A Board reported on the Churn Range, that it was safe; compensation was paid by Government to two of Lord Wantage's tenants, for the periods when they were debarred from using the ground behind the Butts.

The recruits and the various companies proceeded by march route to Churn and back, *via* Pangbourne, Upper Basildon and Aldworth; the distance is 14½ miles, time occupied 4 hours 10 minutes to 4½ hours. The course of practice was a good deal interfered with by rain, and, unless the sight was raised, the black powder used did not carry the bullet to the target.

Major-General Lord Methuen, commanding the Home District, took the opportunity of his Annual Inspection of the Depôt, Royal Berkshire Regiment, to see the regiment.

Colonel Borrett made his Annual Inspection before the end of the training, and expressed his approval of the drill and conduct of the regiment.

Mr. Benyon again placed a portion of his park at the disposal of Colonel Borrett, in order that the regiment might practise the attack before him; but, unfortunately, Colonel Borrett was prevented by illness from being present. The operations were, however, carried out, and the battalion marched back through Theale and the Bath Road to quarters.

The Officers' Luncheon and Sports took place as usual and were very successful, much amusement being caused by an officers' cigar and umbrella race on ponies.

The regiment was in the course of the training, for the first time for some years, exercised in outpost duties, in the neighbourhood of Tilehurst, on a line from the Thames to the Bath Road.

Promotions: Major T. J. Bowles to Lieut.-Colonel, *vice* Lieut.-Colonel Archer Houblon; Captain Otway, Major.

The resignations were: Lieut.-Colonel Archer Houblon (Hon. Colonel), March 20th, 1894. Lieutenants: G. C. T. Willes, April 14th, 1894; S. H. Rickman (to Rifle Brigade), November 29th, 1893; M. M Mercer Adam, August 26th, 1893; C. Knipe, February 3rd, 1894.

Appointments: P. Cazenove, February 10th, 1894; H. G. Henderson, March 3rd, 1894; H. C. F. Hay, March 12th, 1894.

1895.

As has been mentioned previously, the regiment's place in the "Home Defence Scheme" was with the 2nd Army Corps, in the 18th Brigade, with Head-Quarters in the South-eastern District. A small sum of money was available for manœuvres, or tactical exercises, in that district in 1895; and it was determined to mobilise, *i.e.*, make mobile, for the first time, a force of Militia for one week, and at the same time to try the experiment of steam traction, which is thought by some authorities to be the ideal transport for home defence on English roads.

In accordance with this idea, the regiment was ordered to join its brigade in Camp at Lydd, near Dungeness, on May 13th, having been preceded by fourteen days by the recruits and advance party for the recruits' musketry course. Accordingly, after the Assembly and Muster, the 3rd Royal Berkshire entrained at the South-Eastern Station at Reading the same day, and, after six-and-a-half hours' journeying in great heat, marched into Camp on the flats of Lydd, in the dark, at about 9 p.m.

The Brigade Camp was placed just to the south of the Rifle Ranges, opposite to the brick huts of the Royal Artillery. The Royal Sussex on the right, West Kent in the centre, and Royal Berkshire on the left. The remaining battalion, the Royal West Surrey, was at New Romney with one of their own Line battalions. The brigade was under the command of Colonel Tolson, of the 35th Regimental District, from Chichester.

Lydd was formerly on the sea, now it is about two-and-a-half miles inland. The Camp lay on the reclaimed land, with acres and acres of shingle to south and west, and was much exposed to the winds on south-west from the sea and on the east over Romney Marsh.

The work of the training went steadily on, facilitated by the close proximity of the ranges, where Lieut. Swinton, of the Depôt, attached as Instructor of Musketry, put the

regiment through the course. Some outpost work and
battle formation, popularly known as "shingle-punching"
(being carried out on the shingle), was got through by
battalions and afterwards in brigade. The Surrey were
brought in and the brigade was inspected by Lord W.
Seymour, on Monday, May 28th.

Mobilisation took place and the 18th Brigade marched
to a position near Rye, on June 3rd, leaving a party behind,
consisting of the Quarter-Master, one Subaltern, and about
sixty men, who rejoined with the heavy baggage at Ashford
on the following Friday. The 18th Brigade formed part of
a force supposed to have landed near Lydd, at 4 a.m., on
June 3rd. The defending force acted from Ashford as a
base.

The regiment had a very hard day the first day. Packing
of tents, kit-bags, blankets, etc., had to be done before
marching out for the Brigade rendezvous, which was reached
at 8 o'clock. The march was long, the weather very close
and sultry, the roads loose and dusty. The camping ground
was reached in good time, but was not marked out. The
men did not get their dinners till 4 p.m., owing to a partial
breakdown, which specially affected the waggons allotted to
us. The transport only arrived at 10-30, when everything
had to be unloaded and carried some distance from the
roadside into camp. Tents had to be pitched, kit-bags
and blankets sorted out, and the men turned in. Luckily
it was a fine night; officers and men worked with a will,
but when "lights out" sounded, I was afraid there would be
many things missing in the morning ; this, however, turned
out not to be the case. The men were wonderfully cheery
over their difficulties ; the officers were worse off than the
men (the mess waggon not arriving for hours after the men's
dinner was served), and this, of course, the men knew. One
officer, one or two of the non-commissioned officers, and men
suffered slightly from heat apoplexy, but were all right in the

morning. The mess caterer, Captain Lewis, (late Quarter-Master), was seriously ill, but, recovering somewhat, he pluckily continued his work to the end of the training. Colonel Davis and the Officers, Royal West Surrey, whose transport arrived in fair time, were most hospitable, and entertained many of the officers in the evening, in turn, as they could be spared from their duties. As may be imagined, the night was a short one for all ranks; the Company Officers had to be in their lines at 5 o'clock next morning, as they had to be each morning during the mobilisation. The men turned out in capital spirits, struck camp, packed the waggons, breakfasted, and marched at an early hour for the second day's operations of the mimic campaign. During the previous day our brigade had not been actually engaged. There had been some cavalry work done, and a small cavalry engagement, in which cyclist scouts and signallers had also been employed. Near to our camp was a Volunteer Brigade, not mobilised, but in a standing camp; they belonged to the British, or defending force, but were placed where they were for convenience. They took part in most of the operations when they could reach the scene from their camp. Captain Thornton had been appointed transport officer for the brigade, working under Captain Donovan, of the Army Service Corps. (This officer had been through the first Matabele War under Major Forbes).

On the second day of the campaign the 18th Brigade joined hands with the other brigade of the division (Colonel Paton commanding the division), and fought its way another step towards Ashford, the large junction of the South Eastern Railway, and the objective of the invaders. The ground fought over this day was well adapted for the purpose. The defending force which consisted of Regular Infantry, Cavalry, Yeomanry, and Volunteers, were forced back in the direction of their base at Ashford, and we went into camp for the night, everything now and henceforward working

well. The men marched and fought in marching order,
carrying their valises. The operations of the following two
days were much the same; the defenders retiring before
the invaders, as a result of the day's fighting, except that, in
reality, the engagement fought on the fourth day was
adjudged not to have been entirely favourable to the
invaders, but the programme had to be carried out, and we
marched into camp in the outskirts of Ashford. On this
day, in the absence of Colonel Tolson (away inspecting a
battalion in his command), Colonel Lord March commanded
the brigade, and during the decisive action of the day
assumed command of the division on Colonel Paton being
put out of action, supposed to have been wounded. On this
day, and the following, the men were relieved of their valises.

On the fifth day, Friday, June 7th, we, the invaders,
marched out to the position we occupied immediately
previous to the action of the day before, and repeated the
operations of that day, when, owing to a better combination
of the force at our disposal, we were adjudged to have
been successful, and it was thought that the Commander of
the invading force landed at Lydd on the Monday morning,
had succeeded in carrying out his orders, and seized the
town and important railway junction of Ashford. Could
this plan of fighting over again, with the same resources, the
action of the previous day be adopted in real warfare, what
a different history of some campaigns would have to be
written. For introduction purposes, however, it is a very
good object lesson. On the side of the defenders an
armoured train was experimented with.

After the " cease fire " had sounded, the whole of the
forces engaged were formed up and marched past H.R.H.
the Commander-in-Chief on some very good ground close
by. His Royal Highness spoke in complimentary terms of
the marching powers, conduct, and discipline of the Militia
Battalions. Two men only of the Royal Berkshire fell out

during the week. After the march past, the various regiments mached off, the 18th Brigade returning to Camp at Ashford. Lord William Seymour, commanding the South-Eastern District, was in command of the whole of the forces engaged, and all the arrangements for the conduct of the operations were carried out by his Staff.

After Mess on this, the last night of the training, the officers of the various regiments in the Brigade exchanged farewell visits. The whole Brigade had got on very well together, and, though exhibiting a very healthy spirit of rivalry, parted with mutual expressions of regret.

At 3.30 a.m., on Saturday, June 8th, the regiment entrained at Ashford, arrived at Reading about 7.30 and was dismissed the same afternoon.

It is interesting to note that the last time the regiment had been in those parts was when it lay at Rye, in 1793, and at Ashford in 1803.

The regiment, in addition to Lord W. Seymour's inspection in Brigade, was inspected in drill and interior economy by Colonel Tolson, commanding 35th Regimental District (Acting Brigadier). There were no courts martial during the training, and hardly any military crime.

The officers, when not at work, found some difficulty in filling up their time, but visits to the Dungeness Light-house and the Signal Station, trolleying on the Government Railway to the shore, and the ubiquitous "bykes" helped out; the latter were sometimes put to strange uses as steeplechase "gees," in which pastime Bobby (Hargreaves), with a large regalia in full blast, took many a toss with his usual imperturbability,

The men seemed much to enjoy the fresh fish they got direct from the boats, and the sea bathing. The only grumblers, I believe, were the Field Officers, condemned, as we were, though fat and scant of breath, to "shingle punching" in long boots and spurs, instead of "putties."

The regiment was very successful in the Brigade Sports, carrying off both the regimental and officers' tugs of war and many other events, and also winning most of the cricket matches. There was, one evening, a very successful grand torchlight Tattoo by the bands of the Brigade, organised by Captain Hamilton, Brigade Major, at which many civilians were present.

Unquestionably an occasional training in Brigade with other Militia regiments and in contact with Regular troops, tends to a very useful rivalry in drill and general smartness, as well as in all manly sports, and increases *esprit de corps* and a knowledge of their duties in all ranks.

Since the training of 1894, we had lost Major Otway (Hon. Lieut.-Colonel), Lieutenants P. Downes, Hon. R. C. Craven, S. L. Barry (to 10th Hussars); and the following promotions and appointments were gazetted: Lord G. Pratt to Major *vice* Major Otway; Lieutenant C. Hay to Captain *vice* Major Otway; Lieutenant W. B. J. Van de Weyer to Captain *vice* Lord G. Pratt; B. G. Van de Weyer, A. H. Royds, and W. H. Bagot, Second Lieutenants.

After this training the regiment was taken out of the 18th Brigade, 2nd Army Corps, and was assigned to the Portsdown Forts in the defence scheme, a position much more easily reached from Reading. All the mobilisation stores and transport hitherto stored at Reading were consequently removed. Captain Lewis also gave up the office of Mess Caterer, which he had filled for a long time. On the occasion of his remarriage the year before, the officers, by subscription, had presented him with a piece of plate and a purse.

1896. For this year the suggestion was made from Head-Quarters, through the Home District, that the regiment should train at Aldershot, in August and September, for five weeks, to form part of a large force of Militia, in two divisions, ordered to take part in the manœuvres of the year, to be held at and around Aldershot, owing to the failure of the Government Manœuvres Bill to pass. I was requested to make any remarks I wished upon this suggestion ; and, therefore, while stating that the regiment was ready and prepared to train wherever ordered, I submitted that Berkshire was essentially an agricultural county, that the time named usually comprised the period of harvest, and that the 519 men at that moment on the roll classed as agricultural labourers would be deprived of their harvest earnings, and that in some cases their absence would also bear heavily upon the farmers, whereas, this would not apply in the case of Metropolitan regiments or those from mining and manufacturing districts ; moreover, that, if this suggestion was carried out, it would probably have an adverse effect upon recruiting. The authorities eventually ordained that the regiment should train at Reading in May and early June, proceeding by detachments to Churn for musketry, by march route as in 1894. The ground for the camp of the detachment being lent by Lord Wantage, and water provided free of expense to the country.

The recruits, with the usual percentage of old hands and non-commissioned officers, assembled at Reading on April 27th under Captain Turner of the 1st Battalion, the new Adjutant, Captain Thornton, Lieutenant White, Inspector of Musketry, who had lately passed through Hythe, and the newly-joined subalterns, after having been put through their preliminary musketry drills, they were marched to Churn, completed their musketry course and returned to Head-Quarters in time for the assembly of the regiment.

Since the previous training, not only had a new Adjutant been appointed, but Hon. Captain and Quarter-Master Hollyer had retired on pension, and accepted an appointment offered him by Major A. W. Hay; Quarter-Master and Hon. Lieut. T. Brown, from the 1st Battalion, being

Q

appointed in his place, Sergeant-Major Seely remaining at the head of the Staff sergeants. The roll of officers remained the same, except that Lieutenants R. Cazenove (to 6th Dragoons), P. Cazenove, and R. Hargreaves resigned, and the following new officers were appointed: Second Lieutenants Scott, Johnstone, Wadling, and Drummond (the last-named resigned before the training). Captain Birch and Lieut. Rhodes were granted leave of absence.

There is a very good group of officers inserted in this work, taken by Sergeant-Major Beale, of the Depôt. It has been the custom, for many years, to have a group taken annually.

The training was favoured with exceptionally fine weather and, in consequence, the regiment made good progress. Major-General Lord Methuen made a careful inspection of the Battalion on the occasion of his annual inspection of the Depôt Royal Berks Regiment, and Colonel Dickson, C.B., held the annual inspection of the regiment just previous to its being disbanded. The conduct of the men was exemplary, there was no court-martial, and very little ordinary military crime. Altogether the regiment was very well reported on. The men never before lived so well, modern improvements in the matter of messing and canteen arrangements working well, without any extra cost to the country. It is thought that the provision of hot soup, with bread left over from their allowance, in the evening, is a great help, especially to the younger soldiers. All worked hard, and had plenty of amusements in the way of cricket (inter-company matches), football, etc. The annual sports and officers' luncheon were well attended and successful, one or two improvements on previous years being introduced.

A small tactical day's instruction, Major Thoyts' park being kindly lent for the occasion, was arranged to practise out-post work and battle formation, a rear guard action being fought; Major Wheble holding Sulhamstead Park,

GROUP OF OFFICERS, 1896

its approaches, and adjacent woods, as Commander of the rear guard of a force retiring westwards along the southern ridges of the Kennet valley. His opponents, under my command, represented the advance guard of a strong force covering Reading, whose outposts had held during the night the line of the Bath Road, Reading-Newbury Railway, and the river Kennet from Theale to the Reading-Basingstoke line. Major Wheble was outnumbered, and though, according to his orders, holding on to Sulhamstead Park and its coverts as long as possible, was forced to retire with considerable loss. The regiment was then formed up, arms were piled, the men fell out and were fed under the trees in the Park.

Major Thoyts—who was for many years in the regiment, being present with it in Corfu—had invited Colonel Dickson, who was present to inspect and report upon the day's operations, myself, and the Officers of the Battalion, to luncheon in the house ; most unfortunately, owing to the death of a relation, at the last moment neither he nor any of his family were able to be present, but, all arrangements having been made, he would not allow it to be put off, and sent a kind message deputing me to act as his representative for the occasion. After luncheon and a rest for the men, the Attack was practised on the way back to Barracks the reverse way of the Park.

The Park, with its fine timber, was looking at its best ; and the outing was a pleasant change to all ranks from the monotony of the drill field. I believe some pheasants and partridges nesting in the Park remained undisturbed all through. One partridge never left her nest, which was in the gravel pit close to where the men halted for their dinner. Major Thoyts told me afterwards he was surprised at not finding a single cartridge case left upon the ground, showing how carefully the men had obeyed orders and picked them up as they advanced or retired. I believe this was the first time for many years, certainly in my recollection, that the officers of the regiment had been entertained in this way.

Owing to a local outbreak of small-pox, supposed to have been imported from Gloucester, a few men were given leave from the training for fear of infection, and a slight outbreak in Reading necessitated confinement of the men to Barracks for a day or two; but this being energetically dealt with by the Local Sanitary Authority, the danger was soon over. The great improvement in stamina of the men, after a week or two of good food and work, was specially notice-able. There were few absentees.

It may be noted that all Militia recruits, during their recruit training, are now put through a modified course of gymnastics, in addition to their ordinary drills; a matter of great importance to young lads. A new sword exercise for officers was introduced and taught daily.

Several changes of uniform have, in the dress regulations for the Army, been introduced for officers. Brass scabbards for infantry mounted officers have been abolished; brass spurs also, except for Levee dress. The old blue patrol jacket has been superseded by a blue serge fatigue jacket, and the rolled collar mess jacket re-introduced, after many years; it is, undoubtedly, a more comfortable garment. Officers are permitted, however, to wear out dress of old pattern for two or three years. The sword is now ordered to be worn with the belt under the red serge, which has now, for some years, taken the place of the tunic (except in review order) and the blue patrol.

The ladies of the county, under the presidency of Lady Wantage, with a committee of three ladies of the regiment, Lady George Pratt, Mrs. Hay, and Mrs. Thornton, raised a subscription to present new colours to the regiment, in place of the very heavy old pattern ones, also presented on behalf of the ladies of the county on July 30th, 1855, by Lady Abingdon. It was proposed that the presentation should be made during the forthcoming training, but owing to the training taking place away from Head-quarters, at Churn, and the many Jubilee engagements, it has been found necessary to postpone it to 1898.

Since the training of 1896, Major T. J. Wheble, Captain F. M. Birch, and Lieutenant Bagot have resigned. Captain and Hon. Major A. W. Hay has been promoted Major, and Second Lieuts. Purnell, Adams, Chamberlayne, Urlwin, and Archer Houblon have been appointed.

Lieutenant Evans, 3rd Worcester, and Lieutenant Dauncy, 7th Rifle Brigade, were attached for the 1896 training.

1897. The regiment has been ordered to train at Churn in May and June, in brigade with the 3rd and 4th Oxfordshire Light Infantry, though at one time it was feared that the training would take place at Aldershot, for manœuvres, during harvest. Recruiting has not been good.

In conclusion, I think I may venture to hope that the regiment has every prospect of maintaining the high character it has always sustained for discipline, good conduct, and hearty and ungrudging efforts to carry out its duties. The rank and file of Berkshire men make excellent soldiers, are most willing and amenable to discipline.

The history of the regiment during the time I have served in it would hardly be complete without allusion to those who have served in the rank of Sergeant-Major; that connecting link between the Commissioned ranks and the Staff Sergeants and men, and whose smartness and fitness, or otherwise, for their duties, may make so much difference to the efficiency and conduct of the permanent Staff, Non-Comissioned Officers and men. I will venture to hope that some day we may see Militia Sergeant-Majors again with warrant rank.

When I joined, Sergeant-Major Staden, appointed 1855, held the post (which he retained until October, 1876, when Serjeant-Major Butler was appointed), he was a non-

commissioned officer of the old type.   Those who remember
him will, I think, agree with me that a better Sergeant-
Major in every respect than Sergeant-Major Butler, or a
smarter drill, would be almost impossible to find.   Sergeant-
Major Staden came from the 13th Light Infantry, and
Sergeant-Major Butler from the 46th Regiment.   After
them came Sergeant-Major Dunn, promoted in the regiment,
he soon after got an appointment at the Recruiting Depôt
in London.   Then came a short interregnum, when we had
an Acting Sergeant-Major, and then Sergeant-Major Seeley,
who now holds the appointment, was promoted.

With few exceptions, I think, during all this time the
Permanent Staff have done their work well, and with zeal
and efficiency.   As in the case of the Sergeant-Majors, I
should like to see an alteration made in the status of the
Orderly Room Clerks.   An addition, too, of two or three
Sergeants to the Permanent Staff, would be a great help to
Militia battalions, so as to have always two Staff Sergeants
for duty with each company.

. There is a good group of the colours and a colour party,
taken, like the others, by Sergeant-Major Beale.

The Officers' Mess possesses a very fair chest of plate,
and among many pieces presented to the regiment, reference
may perhaps be made to the following and their donors.
Centre-piece: an' elegantly designed piece of plate, presented in 1855 by the ladies of the county ; a large and very
heavy old silver Tankard, capable of holding half-a-dozen
bottles of wine, given by Lord Abingdon (then Lord
Norreys) when his horse, "Sir Bevys," won the Derby in
1879 ; two large and handsome Cups and Covers, copies of
the old Irish Cup, given by Major Hay ; two silver Ink-
stands, one given by Lord Algernon Percy ; two Menu
Stands, by Captain Nepean ; a very handsome Chest of
forty-eight fish knives and forks, by Lieutenant Cazenove ; •

THE COLOURS AND COLOUR-SERGEANTS, 1896.

Sergeant-Major A. SEELEY,
3rd Royal Berks Regt.

Colour-Sergeant F. MAUNE.
3rd Royal Berks Regt.

Colour-Sergeant W. THOMAS.
3rd Royal Berks Regt.

two very handsome cut-glass, silver-mounted, Champagne Jugs, by Lieutenant Barry; a set of Mustard Pots, by Captain Van de Weyer; three handsome Lamps, by Lieut. Hargreaves; an Inkstand, by Major Otway; a handsome Clock, by Captain Adams; two Cigar Lighters, consisting of a copy of the badge of the "Stag under the Oak," by Captain Thornton; and a Silver Bowl, by Lieut. Henderson.

The camp at Churn is already formed and occupied by the recruits, and a proportion of non-commissioned officers and duty men of the 4th Oxford and 3rd Royal Berks for musketry, they having assembled on April 26th. The 3rd Oxford (Bucks Militia) are carrying out their recruit musketry at Wycombe, but will join the Camp when all three regiments assemble on May 10th. Lord Wantage has again granted the authorities the use of the ground and water supply. He has, moreover, lent us the large hut upon the ground for our Mess-room, together with stabling at one of his farms for the mounted officers' horses. Messrs. Unite, of London, have supplied the ante-tent, kitchen, etc., and fitted up the Mess-room in a very satisfactory manner, under the direction of the Mess President, Major Lord George Pratt, who we shall be very sorry to lose, on his retirement after this training. Our own furniture has been brought from Reading. Messrs. Cross and Jameson have been engaged as Messmen. Altogether the Mess promises to be a very comfortable one, at no great expense; having the hut for a Mess-room, thanks to Lord Wantage, is a saving of expense. It has always been the custom in the regiment, while endeavouring to have everything necessary for reasonable comfort, not to unduly increase the expenses of the Mess for the sake, especially, of the young officers.

Colonel Kingscote, commanding 43rd Regimental District (Oxford), has been appointed Brigadier. Probably the Brigade will be inspected, during the last week, by Major-General Lord Methuen, commanding the Home District.

# CHAPTER XIII.

## PLACES WHERE THE REGIMENT HAS BEEN.—1614-1896.

THE soldiers of Berkshire without doubt exercised annually from the earliest period, but of these earliest Musters I have not enquired. Among the Reading Corporation MSS. are references to the calling out of troops during the reign of Queen Elizabeth. Another bundle of MSS. giving the following interesting particulars, being the precepts issued by the Lord Lieutenant or Deputy Lieutenants to the Mayor and Corporation for mustering the Trained Band of Reading, both "horse and foot, clergy as well as laity."

The Trained Bands were called after the locality in which they were raised. They exercised annually in the chief town of their division. Thus we have an Abingdon troop, a Reading troop, also troops from Newbury, Wokingham, the Forest, etc.; so they continued to be called until this century; now the companies are numbered, not named. The whole regiment was known collectively as the Trained Band or Militia of the county.

From the precepts referred to, I am able to give the dates of exercising of the Reading company; no doubt similar precepts were sent at the same time to all the Divisions. Each training occupied two days: the first two being at Whitsuntide, the last two being later in the year. This continued until the reorganization of the Army in 1758.

### READING TROOP.

1614. October 10th.     The Forbury, 8 a.m.
1615. September 11th.   The Forbury, 7 a.m.

| 1617. | September 23rd. | The Forbury, 8 a.m. |
| 1618. | September 18th. | The Forbury. |
| 1619. | May 8th. | The Forbury. |
| 1620. | October 5th. | —, 9 o'clock. |
| 1621. | April 9th. | Bulmershe Heath. |
| 1622. | July 16th. | Bulmershe Heath. |
| 1623. | September 15th. | The Forbury. |
| 1624. | June 21st. | The Forbury. |
| 1625. | October 10th. | Reading, 8 o'clock. |
| 1626. | August 8th. | Bulmershe Heath. |
| | October 9th. | The Forbury, 9 o'clock. |
| 1627. | Whitsun. Week. | The Forbury. |
| 1629. | June 30th. | The Forbury. |
| | July 20th. | The Forbury. |
| 1630. | July 30th. | The Forbury. |
| | September 22nd. | The Forbury. |
| 1631. | July 26th. | The Forbury. |
| | September 5th. | The Forbury. |
| 1632. | May 27th. | The Forbury. |
| 1633. | June 7th. | The Forbury. Reviewed before Sir F. Knollys, in the Forbury, on June 12th. |
| 1634. | May 28th. | The Forbury, 7 o'clock. |
| | July 15th. | The Forbury. |
| | August 9th. | |
| 1636. | June 8th. | The Forbury. They were ordered to be in " Modern Fashion." |
| 1637. | Whitsun. Week. | The Forbury. |
| 1638. | May 16th. | The Forbury, 7 o'clock. |
| | December 16th. | |
| 1639. | October 4th. | The Forbury. |
| 1640. | August 4th. | The Forbury, 8 o'clock. Clergy as well as laity were summoned, and four, five, or six young men in each parish ordered to be ready in case of need. |

1642.   September 1st.     The Forbury.
        October

During the Civil War the regiment moved about where
it was required for defence or fighting.

I have very little doubt, although no record remains
to prove, that the Berkshire Militia—I may here remark
the term Militia was used in reference to the Berkshire
Regiment as early as 1640; the word seems to have first
appeared in Queen Elizabeth's reign—was either embodied
or trained yearly until the year 1667, when we know they
were sent to the Isle of Wight to protect the coast.   During
the Duke of Monmouth's rebellion they were doubtless on
active service.   It is usually asserted that the Militias were
neglected at this period, but I believe they exercised
annually.   From 1758 to the present day I am able to give,
year by year, the localities of the Berkshire Militia.

It is remarkable that although Abingdon was the county
town, it was never used as a military station or training
place by the county Militia, except as a store place for a
few years during the present century.   Reading and
Newbury lay on the main road between London and the
important garrison of Bristol, and were also better placed
for speedy communication with the Southern coasts.

The regiment, when moving from one place to another,
usually marched in two divisions.   They were billeted out
as was found most convenient among the villages.   Thus I
have given the names of every village or place mentioned in
connection with the regiment.   Many weeks must have been
spent on the march.   They, like all other Militias, were moved
about wherever their services were required, but were
principally employed for coast defence.

The plan of defence adopted in 1792 was, as far as possible,
copied from that of Queen Elizabeth's Army in 1585-88, and
I have little doubt that the military arrangements of the 16th

Century were drawn on the plan of a still older model, so that all the camps of the South of England have been used as military positions of value from time immemorial.

### PLACES WHERE THE BERKSHIRE MILITIA HAVE BEEN ENCAMPED, QUARTERED AND TRAINED.

| | | |
|---|---|---|
| 1758. | December. | Embodied. Strength 560, as decided. |
| 1759. | June. | Reading. Review at Whitley Wood. |
| | July. | Marlborough, Hungerford and Devizes. |
| | August 21st. | Newbury for a week, then to Devizes. |
| | August. | Two companies from Hungerford, to Marlborough and Preshute. |
| | October. | Winchester Barracks. |
| 1760. | June 17th. | Winchester Camp. |
| | October 9th, 10th. | Hungerford, Ilsley, Newbury and Speen. |
| | November 5th. | 5 companies to Reading. |
| | | 2 companies to Wallingford. |
| | | 2 companies to Wokingham. |
| | | — |
| | | 9 companies. |
| 1761. | March 1st. | Two companies to Witney, to help quell riots; remained there about a fortnight. |
| | April. | Reading. They were sent out of the towns during the Elections, and were at Abingdon and other places, but returned to Reading and were ordered to Winchester Camp, June 12th and 13th. |

| | | |
|---|---|---|
| 1761. | October 20th. | Left Winchester Camp for Reading Head-Quarters, Wallingford and Wokingham. |
| | November. | Newbury. |
| 1762. | March. | Winchester, to guard French prisoners. |
| | April. | Returned to Reading. |
| | June. | Ordered to Winchester Camp, to arrive on 18th and 19th. This order must have been changed, as they were at Reading on June 21st. |
| | June 23rd. | Ordered to Winchester, to arrive there by 26th inst. |
| | Oct. 24th & 25th. | Left Winchester Camp and returned to Reading. |
| | October 27th. | Wokingham. |
| | | Reading. Disembodied before Christmas. |
| 1763. | | Probably were not trained that year. |
| 1764. | | |
| 1765. | | |
| 1766. | | (?) Newbury. |
| 1767. | | Reading. |
| 1768. | May 2-25th. | Newbury. |
| 1769. | October 4th. | The Forbury, Reading, and Newbury. 28 days. |
| 1770. | May 7th. | Newbury, June 2nd. |
| 1771. | May 6th. | The Forbury, Reading. 28 days. |
| 1772. | May 4th. | Market Place, Newbury. 28 days. |
| 1773. | May 3rd. | The Forbury, Reading. 28 days. |
| 1774. | May 7th. | Market Place, Newbury. 28 days. |
| 1775. | May 8th. | The Forbury, Reading. 28 days. |
| 1776. | May 6th. | Market Place, Newbury. 28 days. |

| | | |
|---|---|---|
| 1777. | May 5th. | Market Place, Reading. 28 days. |
| 1778. | | Reading Head-Quarters. |
| | June 13th. | Coxheath Camp. |
| | November 9th. | Reading. |
| 1779. | | Two companies at Woodstock, February to June 5th; nine companies at Reading, left on June 7th. |
| | June 11th. | Romford, Ilford, Hare Street and Adarley Common. |
| 1780. | May 30th. | Reading. Five companies left for Winchester, to guard prisoners of War; four companies to Hilsea Barracks, started June 10th, to arrive on the 14th. Left Hilsea Barracks, Portsmouth, on October 17th and 18th, for Oxfordshire. Two companies to Banbury and New-thorpe, one company at Burford, one company at Doddington, Adderbury and Bloxham, one company at Witney and Eynsham, two companies at Chipping Norton and Chapel House, one company at Bicester, and one company at Islip and Bletchingdon. |
| 1781. | April 9th. | Moved to billets near London. Four companies to Barnet, Hadley, Kitts End, Ridge Mins, Potters Bar and Northall, one company to Whetstone, three companies to Hampstead, Highgate, Hornsey and St. Pancras, |

|  |  |  |
|---|---|---|
|  |  | one company to Stanmore, Edgware and Bushy. |
| 1781. | May 7th. | Detachment at Paddington. Five companies to Maidstone, two companies to Sevenoaks, Seal and Riverhead, two companies to Wrotham, Ightham, Offham and Mallings. |
|  | June 7th. | Lenham Heath. Nine companies. |
|  | October 20th. | Left Lenham. Two companies to Sevenoaks, Seal, and Riverhead; one company to Tonbridge and Hadlow; three companies to Tonbridge Wells; one company to Lamberhurst, Gondhurst, and Horsemunden; one company to Cranbrook, Milkhouse Street, and Hawkhurst; one company to Westerham, Brasted, and Tonbridge, and other villages adjacent. |
| 1782. | July 1st. | Coxheath Camp. |
|  | November 12th. | Light Company to Rochester and Stroud. |
|  | November 15th. | The remaining companies to Rochester, Stroud, Finsbury, Brompton, and Gillingsham. |
|  | November 28th. | The regiment marched to Newbury, Speen, and Speenhamland, and Oakingham (Wokingham), December 4th. |
| 1783. | March. | Reading. Disembodied. |
| 1784 |  | They were never trained the year following disembodiment. |
| 1785 |  |  |

| | | |
|---|---|---|
| 1786. | | Probably were out for a month's training. |
| 1787. | May 7th. | Newbury. 28 days. |
| 1788. | May 12th. | Reading. 28 days. |
| 1789. | May 4th. | Newbury. 28 days. |
| 1790. | May 10th. | Newbury. 28 days. |
| 1791. | May 16th. | Newbury. 28 days. |
| 1792. | December 18th. | The Forbury, Reading. Embodied. |
| 1793. | | South coast towns. |
| | July. | Left Hastings. Proceeded to Waterdown, near Tunbridge Wells. A detachment, with French prisoners, to Rye. |
| | December. | Four companies to Southampton and Romsey. |
| 1794. | April. | Eastbourne and Hythe. |
| | May. | Deal and Ramsgate. Dover to Sandgate. Dover to |
| 1795. | May. | Shorncliffe, Hythe, Margate and Ramsgate, Sandgate and Isle of Thanet. |
| 1796. | | Totness and adjoining towns. |
| | November 28th. | Plymouth Garrison Dock Lines. |
| 1797. | | Dartmouth and Bristol. |
| 1798. | | Bristol. The Militia Horse Troop was at this time called the Berkshire Provisional Cavalry. It was sent to join the Militia under Major Stead, and from there proceeded to Ireland. Landing at Pigeon House, Dublin, in June. This is the last notice of the Militia Horse. |
| | August. | Poole Barracks. |

| | | |
|---|---|---|
| 1798. | September. | Weymouth. |
| 1799. | November. | Portsea Barracks, Southampton, Netley, and Portsmouth Barracks. |
| 1800. | | - Portsmouth. and Weymouth. |
| 1801. | | Weymouth. |
| 1802. | | Reading. |

The following is taken from the old Court-martial books of the regiment.

| | | |
|---|---|---|
| 1803. | April. | Reading. The Mess-room being at the "Crown" Inn. |
| | June 2nd. | "The Ship" in Faversham, on the march to Ashford *via* Kingsdown, Kent. |
| | June 30th. | Ashford Barracks and Ashford Camp. |
| | October 15th. | Shorncliffe Camp. |
| | December 1st. | Walmer Barracks. Deal Barracks. |
| 1804. | March 5th. | Deal, North Infantry Barracks. The Nore, left October 30th for Chelmsford. |
| | December 28th. | Stoke Barracks, Ipswich. |
| 1805. | August 28th. | Walmer. Men were sent to work at Dover. |
| | October. | Taunton. |
| | December. | Detachment at Taunton. Bridgewater with Prisoners of War. |
| | December 13th. | Berry Head, Torbay and Brixham. A Guard at Fishcombe Battery. |
| 1806. | November 19th. | Portsmouth Barracks. |
| 1807. | February 3rd. | Colwart Barracks and Portsea. |
| | July 7th. | Steyning Barracks. |
| | July 20th. | Blatchingdon Barracks. Men in quarters at Lewes. |
| 1808. | March 1st. | Hailsham Barracks. |

| | | |
|---|---|---|
| 1809. | July | Yarmouth. [No Marching Orders can be found from 1809 to 1815]. |
| 1810. | November 30th. | Norman Cross. |
| 1811. | June 12th. | Weeley Barracks. |
| | November | Nottingham. |
| 1812. | February 12th. | Nottingham, while the Sherwood Foresters were serving in Ireland; from there they were sent to quell the riots at Manchester, Liverpool, and other places. |
| | April 30th. | Three hundred of the regiment passed through Derby. |
| | May 25th. | Preston, Blackburn, and Colne Barracks. |
| | June. | Liverpool. |
| | November. | Sommerton. Bideford. |
| | December 28th. | Bideford to Plymouth and Mill Bay Prison Barracks. |
| 1813. | January 22nd. | Marlborough Barracks. |
| | April 29th. | Sailed from Plymouth for Ireland. Landed at the Cove of Cork on May 11th or 12th, possibly a detachment was left behind at Blackburn. The *Militia Register* states that they embarked May 7th. |
| | | Middleton to Athlone. |
| | August 16th. | Left Athlone for Galway. |
| 1814. | September 21st. | Left Tuam for Newry in North of Ireland, where they were expected September 29th and 30th to embark for England. |

R

| | | |
|---|---|---|
| 1814. | October. | Liverpool.  Returned to Reading and Wokingham |
| 1815. | January 12th. | Domingo House Barracks. |
| | July 15th. | Fort Barracks. |
| | | ✦ Liverpool. |
| | September. | Reading. |
| 1816. | March 14th. | Disembodied all except the permanent Staff. |
| 1817-52. | | The lists of officers were regularly entered in the Army Lists year by year, with all additions and alterations. |
| 1852. | November 11th. | Reading.  Reorganised.  21 days. |
| 1853. | October 20th. | Reading.  28 days. |
| 1854. | May 11th. | Reading.  28 days. |
| 1855. | January 1st. | Embodied.  Sailed for Corfu. |
| | October 17th. | Landed at Corfu. |
| 1856. | May 26th. | Left Corfu. |
| | June 13th. | Reading.  Paid off July 4th. 185 days embodied. |
| 1857. | September 30th. | Embodied at Reading. |
| | October 1st. | North Camp, Aldershot. |
| 1858. | April 25th. | Left Aldershot.  Paid off May 7th.  219 days embodied. |
| 1859. | July 15th. | Reading.  21 days. |
| 1860. | May 4th. | Reading.  Inspected May 28th. 27 days. |
| 1861. | April 22nd. | Reading.  27 days. |
| 1862. | May 8th. | Reading.  Inspected by Colonel Alison, May 27th.  Paid off May 28th.  21 days. |
| 1863. | May 4th. | Reading.  21 days. |
| 1864. | April 21st. | Reading.  Inspected May 10th. Paid off May 11th.  21 days. |

| | | |
|---|---|---|
| 1865. | May 1st. | Reading. Inspected May 26th. Paid off May 27th, 27 days. |
| 1866. | April 23rd. | Reading. Inspected May 18th. Paid off May 19th. 27 days. |
| 1867. | April 29th. | Aldershot. 27 days. |
| 1868. | April 27th. | Reading. Aldershot. Inspected by General Horsford, May 21st. Dismissed 23rd. 27 days. |
| 1869. | April 19th. | Reading. 27 days. |
| 1870. | April 25th. | Aldershot. 3rd Brigade. Inspected May 20th. Dismissed 21st. 27 days. |
| 1871. | May 1st. | Aldershot. 3rd Brigade. 27 days. |
| 1872. | September 30th. | Aldershot. 2nd Brigade. 27 days. |
| 1873. | May 5th. | Aldershot. 3rd Brigade. 27 days. |
| 1874. | May 18th. | Aldershot. 3rd Brigade. 41 days. |
| 1875. | May 10th. | Aldershot. 3rd Brigade. 27 days. |
| 1876. | July 3rd. | Minchinhampton Common. 2nd Brigade, 3rd Division. 27 days. |
| 1877. | June 25th. | Aldershot. 2nd Brigade, 3rd Division. 27 days. |
| 1878. | May 13th. | Reading First year at the Barracks. 27 days. |
| 1879. | May 12th. | Reading. 20 days. |
| 1880. | May 10th. | Reading. 28 days. |
| 1881. | May 16th. | Reading. Title changed to 3rd Battalion. 27 days. |
| 1882. | July 24th. | Aldershot. 28 days. |
| 1883. | May 7th. | Reading. 27 days. |
| 1884. | | No training, owing to small pox in Reading. |
| 1885. | May 18th. | Reading. 27 days. |
| 1886. | May 10th. | Reading. 27 days. |
| 1887. | June 20th. | Aldershot. Cove Plateau, North Camp. 27 days. |

R 2

| 1888. | May 14th. | Reading. 27 days. • |
| 1889. | May 6th. | Reading. 27 days. |
| 1890. | May 5th. | Reading. 27 days. |
| 1891. | May 11th. | Aldershot. Watts' Common. 27 days. |
| 1892. | May 9th. | Reading. 27 days. |
| 1893. | May 1st. | Reading. |
| | May 16th. | Ashdown Forest. 27 days. |
| 1894. | May 7th. | Reading. Churn Rifle Range. Lee Metford Rifles. 27 days. |
| 1895. | May 13th. | Lydd. 27 days. |
| 1896. | May 11th. | Reading. 27 days. |
| 1897. | May 10th. | Churn Camp. |

## CHAPTER XIV.

IT is curious that no history of the county gives a list of the Lords-Lieutenant of this county, for the office was a most important one, as all military matters were under his control. Some historians say Henry VIII. created the appointment, others attribute it to Queen Elizabeth.

The first Lord-Lieutenant of Berkshire whom I have found mentioned is the EARL OF WALLINGFORD, afterwards Earl of Banbury, 1618—1630. He was succeeded by his nephew, at his own special wish, when age and infirmities grew upon him ; the request was written July 8th, 1630.

HENRY RICH, EARL OF HOLLAND. This nobleman sided first with one political party and then with the other, till finally, he was beheaded for high treason March 9th, 1648.

The Militia was placed under a Parliamentary Council during the Commonwealth.

In 1667 JOHN, BARON LOVELACE OF HURLEY, was Lord-Lieutenant of Berkshire. He died 1670. The last of this family, Neville, Baron Lovelace, died 1736.

1715. CHARLES, SECOND DUKE OF ST. ALBANS. Died 1751.

1751. GEORGE, DUKE OF ST. ALBANS. December 15th.

1761. VERE, LORD VERE.

1771. GEORGE, DUKE OF ST. ALBANS.

1786. WILLIAM, LORD CRAVEN.

1791. JACOB, EARL OF RADNOR.

1819. WILLIAM, EARL OF CRAVEN. November 12th.

1826. MONTAGU, FIFTH EARL OF ABINGDON. May 3rd.

1855. MONTAGU, SIXTH EARL OF ABINGDON. February 28th.

1881. GEORGE GRIMSTON, EARL OF CRAVEN. August 11th.

1884. ERNEST AUGUSTUS CHARLES, MARQUIS OF AILESBURY, January 11th.

1886. ROBERT JAMES, BARON WANTAGE. November 10th.

### COLONELS AND LIEUT.-COLONELS COMMANDING THE BERKSHIRE MILITIA—1640-1897.

SIR JACOB ASTLEY, BART.

CHRISTOPHER WHICHCOTE.

ARTHUR EVELYN.

JOHN BLAGRAVE.

RICHARD HAMMOND.

SIR THOMAS DOLMAN, KT.

SIR WILLOUGHBY ASTON, BART.

ARTHUR VANSITTART.

JOHN DODD.

WILLIAM EARL OF CRAVEN.

JACOB EARL OF RADNOR.

RICHARD SELLWOOD.

CHARLES SAXTON.

PENYSTON P. POWNEY.

EDWARD LOVEDEN LOVEDEN.

SIR FRANCIS SYKES, BART.

GEORGE HENRY VANSITTART.

THOMAS RAVENSHAW.

WILLIAM VISCOUNT FOLKESTONE.

John Blagrave.
Charles Bacon.
Adam Blandy.
Lord Norreys (afterwards Earl of Abingdon).
V. W. B. Van de Weyer.
John Blandy Jenkins.
George Eyre (afterwards Archer-Houblon).
Thomas John Bowles.

## CHAPTER XV.

### OFFICERS OF THE BERKSHIRE MILITIA.

THIS portion of the Militia History has been more difficult to compile than any other; it has been gathered together from many sources, but is still far from complete. The Regimental List of Officers only begins in 1803, it is obviously imperfect, and the ages given in it are far from accurate. The Berkshire Militia of the last century was so often embodied as to be on a footing with the Regulars; then, as now, Ensigns joined and after a short service went into the Regulars; their resignation only was gazetted, without stating to what regiments they went.

It is interesting to notice how, in some cases, several generations of families have belonged to the County Militia, and I have endeavoured to trace the officers hoping that their descendants will keep up this old custom. There are so few "Berkshire" books, that I have tried, as far as possible, to make this unique as a record of Berkshire County families; for in the old days only men of certain position were given Commissions. I hope any errors or omissions will be pointed out to me; and where such occur I trust they may be forgiven, when it is remembered that I have had to rely entirely upon my own researches, and that newspapers and books of reference are not always trustworthy. Where I could, I applied personally to relatives or descendants, which has involved nearly as much writing as all the book put together.

I have never found an instance of county men joining the 66th Regiment, which, although it was given the title of Berkshire, had nothing to do with the county, and the first years of its existence were spent on Foreign Service; nor, so far as I know, was it ever quartered in the county whose name it bore, until recent years.

ABINGDON, MONTAGU BERTIE, FIFTH EARL OF, Lord-Lieutenant of Berkshire, 1826-1854; born, 1781; second son of Willoughby Bertie, fourth Earl of Abingdon; succeeded to the title 1799.

ABINGDON, MONTAGU BERTIE, SIXTH EARL OF, Lord-Lieutenant of Berkshire, 1855-1881.

ABINGDON, MONTAGU ARTHUR BERTIE, SEVENTH EARL OF. Entered the Militia as Lieutenant Lord Norreys, March 12th, 1858; resigned, but was appointed Major by his father, the Lord-Lieutenant of Berkshire, 1861; Lieut.-Colonel, July 31st, 1863; Hon. Colonel, October 27th, 1880.

ACLAND, JOHN FORTESCUE: Ensign and Surgeon, 1762; Lieutenant, 1779.

ADAMS, JOSEPH WILLIAM RICHARD, of Winkfield: Captain, July 24th, 1880; Hon. Major, March 19th, 1870; went to the 4th Middlesex Regiment, August, 1890.

ADAM, MAUGHAN MERCER MERCER: born 1861; Second Lieutenant, March 4th, 1891; Lieutenant, June 4th, 1892; resigned August 26th, 1893. Afterwards he left the Militia, and gave public entertainments in the style of Corney Grain. He was renting a cottage at Cranbourne when he joined the regiment.

ADAMS, FERGUS EUSTACE, of Cannon Hill, Maidenhead: born, November 26th, 1878, at Wick House, Brislington, County Somerset; son of Henry Adams, of Cannon Hill, Maidenhead, by Eleanor, daughter of the Rev. John Fox, M.A., of St. Bees, Cumberland, grandniece of the Rev. John Fox, D.D., Queen's College, Oxford; Second Lieut., April, 1897.

ST. ALBANS, CHARLES BEAUCLERK, SECOND DUKE OF: Lord-Lieutenant of Berkshire; died 1751. They had a house in Windsor, which had been given to Nell Gwyn by Charles II.; but the family place was near Hanworth, County Middlesex.

ST. ALBANS, GEORGE BEAUCLERK, THIRD DUKE OF: Lord-Lieutenant of Berkshire from July 3rd, 1776. Married, 1752, "the Duke of St. Albans to Miss Roberts" (*Universal Magazine*). He died 1787.

The first Duke of St. Albans was the illegitimate son of Charles II., by Nell Gwyn. He was created Baron Hedington, Earl of Burford, and lastly, Duke of St. Albans, in 1684. Their coat-of-arms was the arms of England with the bar sinister. These were painted on the drums of the Berkshire Militia in 1759 and also borne on the colours.

ALDER, JAMES WALKINSHAW BELL: Assistant Surgeon, 1855; resigned December 28th, 1858.

ALEXANDER, HON. CHARLES: born, January 26th, 1854, third son of James Duc Pre, third Earl of Caledon, by Lady Jane Fredrica Grimston, fourth daughter of James Walter, first Earl of Verulam; Captain Royal Tyrone Fusiliers; Captain Berks Militia, March, 1897. By his own wish this was cancelled a fortnight later, and on April 23rd he was gazetted Captain in the 3rd Battalion Norfolk Regiment.

ALLFREY, MOWBRAY, of Stanbury, resides at Flore Field, Weedon: born, 1853; eldest son of Frederick Allfrey, of Stanbury, by Emily, daughter of Sir Robert Mowbray, of Cockairine, Fife, N.B.; Lieutenant, August 3rd, 1872; went to 15th Hussars, December 2nd, 1874; Adjutant Cheshire Yeomanry, 1883-8; retired from the Army, 1888; married, 1881, Hon. Beatrice Augusta Emmeline, eldest daughter of Baron Saye and Sele.

ALLFREY, FREDERICK VERE, of Stanbury, now living at Ashridge Wood, near Wokingham: born, at Binfield, December 21st, 1854; second son of Frederick Allfrey, of Stanbury; Second Lieutenant, March 26th, 1879; Lieutenant, May 10th, 1880; resigned, February 6th, 1884:

had a cattle range for four years at N.W. Terrioz, Canada ; married, 1889, Miss Maud Hamilton Bruce, whose father rented Shinfield Lodge and afterwards Bulmershe.

ALLIN, JOHN W., of East Hendred : born, 1848 ; son of John Allin, of East Hendred, by Henrietta Jane, daughter of J. H. Grieve, of Wandsworth ; Lieutenant, 78th High-landers ; Captain, May 12th, 1875 ; resigned, September 25th, 1881.

ANDREWS : Captain, Forest Division of Trained Bands, 1640.

ANDREWS, SIR JOSEPH, BART., of Shaw : born, 1726, son of Joseph Andrews (who bought Shaw, 1749), by his first wife Elizabeth, daughter of Samuel Beard ; Captain, 1757 ; Major, 1763 ; created a Baronet 1766. He married, 1762, Elizabeth, daughter of Richard Phillips, of Tarrington, County of Herts, but left no issue. Sir Joseph Andrews was the last of the family, he died, December, 1880, univer-sally regretted by all classes of society. His portrait, in the uniform of Major of the Berkshire Militia, I am able to reproduce here, through the kindness of Henry Eyre, Esq., of Shaw.

ANDREWS, JAMES PETTIT, of the Grove, Donnington : born, at Shaw, 1738; half-brother of the above, only son of Joseph Andrews, of Shaw, by his second wife ; Ensign, 1757 ; Police Commissioner for Queen's Square and St. Margaret's, Westminster ; married Anne, daughter of Rev. Thomas Penrose, Rector of Newbury ; died, August, 1797, buried at Shaw. He was author of a *History of Great Britain connected with the Chronology of Europe*, and other works.

ANDREWS, WILLIAM, of Shaw Mill : Captain, 1757 ; • Churchwarden of Shaw-cum-Donnington, 1754-57. . He was, probably, a cousin of Sir Joseph Andrews, as only men of position and means were given commissions ; perhaps he was the Mr. Andrews, of Reading, who died May, 1784. It may have.been his son who owned the Mapledurham Weir Mills, and married, 1783, Annabella, daughter of John Deane.

ANNESLEY, FRANCIS: son of Martin Annesley, D.D., by the niece of Sir John Cotton, Bart.; Ensign, 1712; M.P. for Reading, 1772-80-84 (refused election to Parliament, through ill-health, in December, 1807). He was Master of Downing College, Cambridge, and Hereditary Trustee of the British Museum. Died, April 17th, 1812.

APTHORP, KENDAL PRETYMAN, of Earley: born at Wolston, County Warwick, May 9th, 1861; second son of Captain Richard Pretyman Apthorp (who lived at Bellevue, Earley, and lately at Sonning), by Emma, daughter of Sir Thomas MacMahon. Second Lieutenant, March 5th, 1879; Lieutenant, May 10th, 1880; went to the 18th Royal Irish Foot, April 23rd, 1881; in India, 1882; served in the Black Mountain Expedition, 1888; also in the Nile Expedition with Royal Irish Regiment, 1884-85; Aide-de-Camp to Sir James Lyle, Governor of the Punjaub Station; Staff Officer at Lucknow, and at present Adjutant of the Oudh Volunteers at Lucknow.

ARBUTHNOTT, ROBERT CHRISTOPHER: born, 1830; Lieutenant, February 26th, 1864; Captain, May 4th, 1871; married Miss Brisco, whose father rented Southcote Manor. Major Arbuthnott afterwards lived in the Bath Road, Reading. He went blind, and died in 1889.

ARCHER-HOUBLON, GEORGE BRAMSTONE EYRE, of Welford: born, 1843; only son of Charles Eyre (formerly Archer-Houblon of Welford, who died 1886); Lieutenant, May 4th, 1871; Captain, February 5th, 1873; Major, May 22nd, 1886; Lieutenant-Colonel, June 16th, 1888; resigned March 28th, 1894, with the rank of Colonel; married Lady Alice, daughter of Alexander, 25th Earl of Crawford; took the name of Archer-Houblon with the property of Hallingbury, County Essex, 1893.

ARCHER - HOUBLON, HENRY LINDSEY: born 1877; eldest son of George Bramston Eyre, of Welford Park, who afterwards took the name of Archer-Houblon with the Hallingbury property; 2nd Lieutenant, April 1897.

ASHBROOK, VISCOUNT WILLIAM FLOWER, of Shelling-ford: born 1767; Ensign, June 15th, 1787; Lieutenant, May 4th, 1789; Captain, February 22nd, 1792; resigned, 1795; succeeded to the title, 1780; died, unmarried, at Wadley House, when the title devolved on his brother, the Hon. Henry Flower. His monument is in Shellingford Church.

ASHHURST, CHARLES HENRY, of Waterstock, County Oxon: born, 1856; second son of John Henry Ashhurst, of Waterstock, by Elizabeth, daughter of Thomas Duffield, of Marcham Park; Sub-Lieutenant, June 25th, 1877; transferred to the Carnarvon Rifles, March 19th, 1878; subsequently joined the Royal Sussex Regiment; married Miss Narren in 1896.

ASTLEY, SIR JACOB, BART.: Colonel of the Berkshire Militia, 1640. Created Baron Astley of Reading, 1644. When the soldiers disbanded themselves at Daventry he remained with the Royal Army; he was made Major-General under the Earl of Lindsey, and was in all the battles of the Civil War. Married Agnes Imple, a German lady of family. He died 1651.

ASTON, SIR WILLOUGHBY, BART., of Wadley, descended from Sir A. Aston, Governor of Reading; Colonel of the Berkshire Militia, 1758; married Elizabeth, daughter of . Henry Pye, of Farringdon. In 1764 he sold the lease of Wadley, and died at Bath, 1772.

ATKINSON, CHARLES: born, 1791; Ensign, September 24th, 1811, Lieutenant, November 17th; went to the 14th Foot, December 23rd, 1811.

AUSTIN, JOHN: born, 1782; Lieutenant, December 17th, 1808; resigned, July 20th, 1809. In the Army List of that year there are two John Austins, one a Captain in the 25th Foot (the King's Own Borderers), August 4th, 1809, the other promoted from Major to Captain in 58th Regiment, November, 1809. Mrs. Austen, of Weston, in Boxford, died, 1822.

BACON, CHARLES, of Elcott : born, 1799; son of Anthony Bacon, who rented Benham, and afterwards Elcott; Captain, August 8th, 1822 ; Lieut.-Colonel, November, 1842; Colonel, April 19th, 1861 ; married, June 11th, 1825. By his first wife Caroline, daughter of Henry Davidson, of Cavendish Square, he left two sons, Charles and George William ; he is buried in the family vault at Shaw.

BACON, GEORGE WILLIAM : born, 1832 ; second son of Colonel Charles Bacon ; Ensign, August 9th, 1853 ; Lieutenant, January 16th, 1855 ; resigned, November 21st, 1855.

BABER, THOMAS DRAPER, of Sunninghill Park : grandson of Sir Thomas Draper ; Captain, 1759 ; died at Newton, Cambridgeshire, May, 1783. The property was sold 1769.

BADCOCK, NICHOLAS: Captain, F Troop Berkshire Militia, 1651.

BAGOT, WILLIAM HUGH N., of Wargrave : born, 1877 ; Lieutenant, March 1st, 1895 ; resigned, November, 1896.

BAILEY [or BAYLEY], BENJAMIN, of Caversham : born, 1763 ; Ensign, April 17th, 1798 ; Lieutenant, July 6th, 1798 ; resigned, April 28th, 1806. Mary, wife of Benjamin Bailey, of Dalby Terrace, died in 1825, age 37.

BAKER, JAMES (perhaps of Streatley) : Lieutenant, 1781 ; resigned, 1792.

BARDSLEY, JAMES, of Southampton : born, 1751 ; Ensign, June 9th, 1794 ; Lieutenant, October 18th, 1794 ; resigned, June 5th, 1809.

BARKER, GEORGE WILLIAM, of Stanlake : born, 1832 ; Ensign, May 20th, 1853 ; Lieutenant, January 16th, 1855.

BARKER, FREDERICK GEORGE, of Stanlake : eldest son of the Rev. Alfred Barker, of St. Leonard's, Sherfield ;

Second Lieutenant, March 19th, 1887; Lieutenant, November 17th, 1888; Captain, March 23rd, 1891; married, 1895, Lucy, daughter of —— Harrison, of Bramley, Hants.

BARRY, HON. AUGUSTUS: brother of Richard, Seventh Earl of Barrymore; Ensign, May 22nd, 1790; resigned, 1793. He became a clergyman, and died, December, 1818, buried at Wargrave. ·

BARRY, STANLEY LEONARD, of Windsor: born, 1874; son of Francis Tress Barry, of St. Leonard's Hill, Windsor; Second Lieutenant, 1891; Lieutenant, April 4th, 1894; went to 10th Hussars, June 2nd, 1894.

BARRYMORE, SEVENTH EARL, VISCOUNT BUTTEVANT AND BARON BARRY, RICHARD BARRY, of Wargrave: born, August 14th, 1769; succeeded to the title, 1773; sent to Eton at 14 years of age; Ensign, July 1st, 1789; Lieutenant, May 24th, 1790; died, 1793; buried at Wargrave, February 17th, 1793. The Right Hon. Lord Barrymore conducting sixteen French prisoners from Rye to Dover by the Berkshire Militia under his command, the whole party halted at the turnpike at the top of Folkestone Hill; after taking some refreshment, on regaining his seat in his curricle, a fusee, which he carried with him for the purpose of shooting sea gulls, went off, and shot him through the head. He died in a few moments; he was only 23 years of age, and so finished a short, foolish, and dissipated life, which had passed very discreditably to his rank as a peer, and still more so as a member of society. Another magazine gives the same account, except adding that he was Member for Heytesbury and an officer in the 2nd or Queen's Regiment. He must have joined the Berkshire Militia after the Guards, as he was Lieutenant in the Militia at the time of his death. He represented Heytesbury in Parliament, and unsuccessfully contested Reading. In spite of his dissipated life he was most popular; he also kept a pack of hounds.

BARLOW, THOMAS ARTHUR PRATT, of the Cottage, Sonning, afterwards of Wellbank, Taplow: born, 1821; Lieutenant, August 16th, 1861; Captain, April 19th, 1866; resigned, May 16th, 1874; Proctor in Doctor's Commons; married, May 3rd, 1859, Maria, daughter of Rev. T. A. Powys, of Medmenham, Bucks; died, 1890.

BARLOW, FREDERICK BARRINGTON PRATT: born at Sonning, 1863; second son of the above; Lieutenant, February 11th, 1882; posted to 1st Battalion, Dorsetshire Regiment, May 21st, 1885; transferred to 2nd Battalion Dorsetshire Regiment, March, 1887; served as Adjutant of 2nd Battalion Dorsetshire Regiment from March, 1892, to March, 1896; promoted Captain, May, 1892; served in the Frontier Field Force, in Egypt, in 1885-86, with 1st Battalion; retransferred to 1st Battalion Dorsetshire Regiment, October, 1896.

BARKSTEAD: Colonel, 1645. His name is also spelt Bacster or Baxster. A Parliamentary soldier. Governor of Reading Garrison. Sir John Barkstead, Kt., Lieutenant of the Tower; M.P. for Reading in Cromwell's last Parliament.

BASSETT, FRANCIS: Captain, 1624; probably one of the Bassetts of Tehidy, County Cornwall. Many of the Bassett family fought for the King; they were all staunch Royalists. The family owned land at Drayton, Shaw, and Letcombe Basset.

BATSON, STANLAKE, of Winkfield Place: son of Stanlake Batson, High Sheriff of Berkshire, 1772, who died, 1812, in his 85th year; Ensign, 1792; Lieutenant, March 14th, 1793; resigned, 1795.

BAYNTUN, HENRY: born, 1836; Lieutenant, June 20th, 1855; removed, December 5th, 1858. His father was in the Navy.

BEALES, BENJAMIN: born, 1778; Adjutant, March 30th, 1811; resigned, July 23rd, 1812; Lieutenant, Royal Marines, half-pay, 1812.

BEDWARDS, TOM B.: Lieutenant, January 26th, 1800.

BELLAS, JOSEPH HENRY: most likely a son of Dr. Bellas, Rector of Basildon and Ashamstead, 1767: Ensign, 1779; Lieutenant, 1781; resigned, February 13th, 1787. The living of Yattendon was held by the Rev. George Bellas in 1574; they evidently belonged to that part of the county.

BELLOES [or BELLAIRS]: Captain, 1640.

BERTIE, HON. CHARLES CLAUDE, of Wytham Abbey; born, 1851; fifth son of Montagu Bertie, sixth Earl of Abingdon; Lieutenant, February 23rd, 1871; went to 47th Foot, 1873.

BERTIE, HON. MONTAGU CHARLES FRANCIS (now LORD NORREYS): born, 1860; eldest son of Montagu Arthur, seventh Earl of Abingdon; Second Lieutenant, December 10th, 1877; Lieutenant, June 25th, 1879; Captain, March 10th, 1883; resigned, April 9th, 1886.

BEVER, SAMUEL JOHN, of Wokingham: born, 1779; Ensign, February 25th, 1807; Lieutenant, June 12th, 1807; Captain, December 12th, 1807. Dr. Bever was rector of Mortimer early in this century, and a member of the Aldermaston Bowling Club, 1758; this may have been his son or grandson.

BIRCH, FRANCIS MILDRED, of Rickmansworth: son of John William Birch, of Rickmansworth Park, County Herts, by Julia, daughter of Joseph Arden, of Rickmansworth Park; Second Lieutenant, May 4th, 1880; Lieutenant, July 1st, 1881; Captain, June 12th, 1886; Honorary Major, March 19th, 1890; served in 1895 with 2nd Battalion at Chatham and in Ireland; resigned, January, 1897. His brother, William Henry Birch, inherited the Grove, Old Windsor, from Miss Thackeray.

BIRNIE, JAMES: Lieutenant, 1798. The name Birnie is Scotch. Perhaps son of Sir Richard Birnie, a Bow Street magistrate, who was knighted, and died, 1832, he was in

S

the Royal Westminster Volunteers, and was of humble parentage from Banffshire. *See Armorial Register*, 1832.

BLACKSTONE, HENRY, of the Priory, Wallingford : born, 1763 ; son of the celebrated lawyer, Sir William Blackstone, of Wallingford, who wrote "Commentaries on the Laws of England," by Sarah, daughter of James Clitheroe, of Boston House, County Middlesex ; Ensign, April 24th, 1786; Lieutenant, February 13th, 1787 ; Captain, October 19th, 1792 ; resigned, 1793 ; died, 1826, unmarried.

BLAGRAVE, JOHN, of Reading ; made Major to take command of the Berkshire detachment of horse under Colonel Dalbier, at the Battle of Newbury, 1642 ; married Hester, daughter of William Gore, of Barrow, Surrey; was M.P. for Reading, 1660-79-80-81 ; died, March 3rd, 1703. This was, probably, the son of Anthony Blagrave, by Dorothy, daughter of Sir Thomas Dolman, of Shaw.

BLAGRAVE, JOHN, of Southcote : son of Anthony Blagrave, of Reading ; Captain, 1758 ; High Sheriff of Berkshire, 1762 ; married, 1745, at St. George's, Hanover Square, Ann, heir of Sir George Cobb, Bart., of Adderbury. He bought Calcot from Benjamin Child, 1759. John Blagrave was buried at St. Mary's, Reading, December 17th, 1787.

BLAGRAVE, JOHN, probably of Watchfield: Lieutenant, 1778 ; Sheriff of Berkshire, 1792 ; married Frances, daughter of Anthony Blagrave.

BLAGRAVE, JOHN, of Southcote and Calcot : born, 1781 (according to one authority, and according to the Militia Register in 1772, but that would make him out to have been eighty at the time the regiment was reorganised in 1852) ; son of John Blagrave, of Watchfield, by Frances, daughter and co-heir of Anthony Blagrave, of Calcot, Southcote ; Ensign, June 21st, 1800 ; Captain, March 18th, 1801 ; Major, September 11th, 1803 ; superseded, June, 1809 ; he was, apparently, reinstated as Lieut.-Colonel, July 14th, 1817 ; Colonel, November, 1842 ; resigned in 1861.

BLAGRAVE, JOHN CHARLES: born, 1862; Lieutenant, January 31st, 1882; resigned, November 1st, 1883 (never joined).

BLAGRAVE, JOSEPH: Lieutenant, 1779; Captain, May 25th, 1780; Captain Commissary of the Royal Regiment of Artillery, 1794; resigned, 1795; married, "St. George's, Hanover Square, May, 1782, Captain Blagrave, of the Berkshire Militia, to Lady Leith, of George Street, Hanover Square."

BLAGRAVE, THOMAS, of Lambourn: Captain-Lieutenant, 1759 (the post of Captain-Lieutenant or Colonel-Lieutenant was abolished in 1802); married Catherine, heiress of Charles Garrard, of Kingswood. They had an only son, John Blagrave, who married Frances, heiress of Anthony Blagrave (see above). Thomas Blagrave, buried at St. Mary's, Reading, 1765.

BLAGRAVE, EDWARD, of Magdalen College, Oxford: born, 1794; eighth son of John Blagrave, of Watchfield, by Frances, daughter of Anthony Blagrave; Captain, August 6th, 1845.

BLAKE, JOHN CARTWRIGHT: Ensign, 1782.

BLANDFORD, GEORGE, MARQUIS OF, of Whiteknights: Lieut.-Colonel of Local Militia, 1807, called the 3rd Battalion or Queen's Regiment, it consisted of seven companies.

BLANDY, ADAM, of Kingston Bagpuize: born, 1782; Ensign, January 1st, 1805; Lieutenant, April 5th, 1805; Captain, July 30th, 1805; resigned, November 19th, 1807; married, 1807, Sarah, daughter of William Mott, of the Close, Lichfield, and Wall, Staffordshire; died, October 24th, 1841.

BLANDY, ADAM, of Earley, late Carabiniers and 15th Light Dragoons: younger son of the above Adam Blandy; Captain, October 4th, 1852; Major, September 30th, 1855; Lieut.-Colonel, April 6th, 1861; Chief Constable of Berkshire, 1863; married, 1864, Anne, daughter of Robert Liston, Esq., and widow of A. Dalrymple, Esq., of Norwich. He had two sons, both died young.

BLYTH, CARLETON V.: born, 1851; Lieutenant, April 26th, 1873; Captain, June 9th, 1877; resigned, March 13th, 1880.

BOOTH, WILLIAM: born, 1785; Ensign, February 1st, 1811; Lieutenant, August 1st, 1811; went to the 14th Foot, December 26th, 1814.

BOWLES, FRANCIS RICHARD, of Milton Hill, now living at 38 Belgrave Road, S.W.: born, March, 1830; son of Thomas Bowles, of Milton Hill, by Hester Sophia, daughter of Samuel Sellwood, Abbey House, Abingdon; Lieutenant, November 2nd, 1852; Captain, September 21st, 1853; resigned, October, 1857; married, Louisa, daughter of Rev. Wildman Yates, Vicar of St. Mary's, Reading.

BOWLES, THOMAS JOHN, of Milton Hill and Streatley: born, October 5th, 1852; eldest son of John Samuel Bowles, of Milton Hill, by Mary, daughter of Rev. Ashhurst Gilbert, Bishop of Chichester; educated at Eton and Christ Church; Barrister of the Inner Temple; Lieutenant, December 31st, 1873; Captain, February 15th, 1879; Major, August 18th, 1886; Lieut.-Colonel, April 28th, 1804.

BOWYER, SIR GEORGE, Bart., of Radley: born, 1780, according to Militia Register; 1783 according to Burke; eldest son of Admiral George Bowyer, Admiral of the Blue (who was created a baronet for his bravery in the victory of the Camperdown, 1794); he came of age in 1804, and all the tenants and people around Radley were lavishly enter-tained for the occasion; Captain, May 16th, 1803; resigned March 13th, 1804; married, 1808, Anne Hammond, daughter of Captain Sir Andrew Snape Douglas. He was M.P. for Malmesbury and Abingdon. Died at Dresden, July 1st, 1804.

BOULT, JOHN, of Charridge in Winkfield: Ensign, 1759. Zechariah Boult owned a manor in Binfield. John Boult, Mayor of Maidenhead, 1770. John Boult, eldest son of Mr. John Boult, died at Boston, U.S.A., in 1822.

Died, at the house of his sister, Mrs. Smith, at Maidenhead Thicket, Mr. Boult, of Hawthorne Hill, in his 80th year, 1833.

BOUVERIE, HON. PHILIP PLEYDELL: born, 1777 (Burke says 1788); fifth son of Jacob, second Earl of Radnor; Ensign, August 13th, 1803; Captain, September 27th, 1803; resigned December 5th, 1804; married, 1811, Maria, third daughter of Sir William Pierce Ash A'Court, Bart.

He was a banker in Westminster, and formerly M.P.

BOUVERIE, HON. EDWARD PLEYDELL: born, 1818; second son of William, third Earl of Radnor; Captain, February 23rd, 1838; married, 1842, Elizabeth Anne, daughter of General Robert Balfour, of Balbirnie; Barrister-at-Law, Under Secretary Home Department, 1850-52; Vice-President of the Board of Trade and President of Poor Law Board. He lived at East Lavington Manor.

BOUVERIE, HON. MARK: born, 1851: fifth son of Jacob, fourth Earl of Radnor; Lieutenant, July 4th, 1870; resigned, July 18th, 1872. He was a Barrister.

BRAHAM, WARD SOANE, of London: born, April 29th, 1824; youngest son of Mr. John Braham, the great English tenor; Ensign, May 2nd, 1855; Lieutenant, October 10th, 1857; Captain, February 19th, 1863; Instructor of Musketry. His brother-in-law, Lord Carlingford, writes: "He died, February 26th, 1877, at a cottage that he had owned for some years at View Island, Reading. Having been first buried in Kensal Green Cemetery, his body was removed in 1879, after the death of his sister, Frances Countess Waldegrave, and in obedience to her directions, to the churchyard of Chewton-Mendip, Somerset, where it lies

beside her own.　Ward Braham took a great interest in the
regiment and in his own work as Musketry Instructor.　He
was a very intelligent man and a most agreeable companion,
indeed the most amusing man I have ever known ; at the
same time, so full of good feeling and good taste, that he
never wounded or gave offence in using his extraordinary
powers of mimicry, and indulging a gift for fun and non-
sense that amounted to genius.　Among other things, he
was a born comic actor and irresistibly funny on the stage,
invaluable, therefore, in private theatricals.　There is a
reference to Ward Braham in an epilogue, written by the
late Ralph Bernal Osborne (given in his privately printed
life), when he was actor and stage manager at the Chief
Secretary's Lodge, Dublin—

> ' To him the credit for this night's success—
> He planned alike our scenery and dress !
> A ladies' man ! tho' at rehearsals sage
> He reigns alone, the Atlas of our stage !
> Prompter ! and painter ! ever near at hand,
> To rouge a cheek, or dance a Saraband.'

A more serious tribute to his qualities is to be found in
a letter written by the late Abraham Hayward, to Lady
Waldegrave (see the *Hayward Letters*, 2, 279.　Murray,
1886), in which he says : ' Your brother's death would have
caused me deep regret on your account, had I regarded him
only as an agreeable acquaintance.　But I had formed a
high estimate of his qualities of head and heart.　His fine
and varied humour, in particular, was the result or product
of observation and reflection.　I have had many serious
conversations with him, and I know few men whose advice
or opinion I should have more prized on matters of conduct,
or right feeling in society.' "

BRICKMANN, CHRISTOPHER DEAKE, of Bath : born, 1826 ;
Lieutenant, October 5th, 1852 ; Captain, August 22nd, 1855 ;

Acting-Adjutant, 1858; retired as Major, April 12th, 1871; resigned, May 4th, 1871.

BRISTOW, HENRY: born, at Eling, Hants, February 19th, 1786; son of William Bristow, R.N., by Mary, daughter of Anthony Sawyer, of Heywood Lodge, White Waltham, Berks; Ensign, May 25th, 1805; resigned, September 20th, 1805; Sub-Lieutenant, 1st Life Guards, 1805; Captain, 1st Life Guards, 1808; Major, 11th Infantry, January 7th, 1814; Lieut.-Colonel in the Army, 1830; received a 1st Certificate as Student of Senior Department of the Royal Military College, 1810; served in the Walcheren Expedition, 1809; Gibraltar, January, 1811; one of the officers of the Quarter-Master-General's Staff in the Peninsular Army, under the Duke of Wellington, during a great part of the War; present with Lieut.-General Sir R. Wilson at the bombardment of Cadiz by the French, 1823; married Elizabeth Alchorne, of the Kentish family of Alchorne. This information is kindly given by Miss Bristow, of Broxmore Park, Romsey.

BROCAS, BERNARD: Captain; killed on the King's side in the first Battle of Newbury.

BROCAS, BERNARD, of Reading: son of Bernard Brocas, of Wokefield, who died in 1777; Ensign; Lieutenant, May 20th, 1790; Captain, November 29th, 1793; married, 1769, Miss Hunter, of Beech Hill.

BROCAS, BERNARD, of Wokefield: born, 1802; son of Bernard Brocas, of St. James's, Westminster, and in 1800, married, at St. George's, Hanover Square, Sarah Ann Redhead, of St. George's parish; Ensign; Lieut.; Captain, August 24th, 1826; resigned, February 18th, 1831. He sold Wokefield. The family owned Beaurepaire, in Hampshire. Bernard Brocas collected a very large quantity of ancient armour; it was sold, a few years before his death, at the Queen's Bazaar, in Oxford Street. He died, August 5th, 1839, at Naples.

BROMLEY, HON. HENRY, of Caversham: only son of
Lord Montford ; Ensign, June, 1798 ; Lieutenant, June 29th,
1798 ; Captain, February 4th, 1799 ; gazetted Major and
Lieutenant Colonel in 26th Foot, 1803 ; married, 1793, Miss
Eliza Watts, of Islington.

BROOKMAN, WILLIAM LAWRENCE: Ensign, 1798 ; Lieut.,
December 13th, 1798. Mr. Brookman, of Reading, gave
£200 to the Grammar School. Dr. Brookman, a friend of
Dr. Penrose, of Newbury, married Miss Patty Head, whose
sister, Sally Head, married George Vincent, Esq., of
Thatcham.

BROWN, JOHN: Ensign, January 23rd, 1809; volunteered
to the 20th Foot, April 14th, 1809. There was a Sir John
Brown, Bart., who died at Sunninghill, 1775.

BROWN, TOM : Quarter-Master, September, 1895 ; from
1st Battalion.

BROWNE, RICHARD : Parliamentary Governor of Abingdon, 1644. He displeased his party in 1648 and was
imprisoned in Windsor Castle. He was a citizen of London.
In 1660 he commanded the City of London Militia, and
was Lord Mayor of London. At the Restoration, he was
made Resident in Paris and created a Baronet.

BRUDENELL-BRUCE, LORD BRUCE, HON. CHARLES :
second Earl of Ailesbury ; born, 1773 ; eldest son of Hon.
Thomas Brudenell, second Baron Bruce, of Kinloss, by
Susanna, daughter of Henry Hoare, of Stourhead, County
Wilts, and relict of Viscount Dungarvan ; Ensign, March
28th, 1792 ; married twice: 1793, Henrietta Maria, daughter
of Noel, first Lord Berwick ; 1833, Maria, youngest daughter
of Hon. Charles Tollemache ; created Viscount Savernake,
Earl Bruce, and Marquess of Ailesbury, 1821 ; died, January
4th, 1856.

BRUMMELL, WILLIAM: died 1770; was a confidential
servant of Mr. Charles Monson, brother of the first Lord
Monson. He occupied a house in Bury Street, where

apartments were taken by Charles Jenkinson, first Earl of Liverpool. His son, William Brummell, an intelligent boy, acted for some time as Mr. Jenkinson's amanuensis; was in 1763 appointed to a clerkship in the Treasury, and during the whole administration from 1770 to 1782 was private secretary to Lord North, by whose favour he received several lucrative appointments. Among other things he was agent to the Royal Berkshire Militia, the Chelsea Pensioners, etc., etc. He further increased his means by his marriage with Miss Richardson, daughter of the keeper of the lottery office, and died in 1794, leaving £65,000 to be divided equally among his three children, two sons, William and George, and a daughter. George Bryan Brummell, known as Beau Brummell, the younger son, was born June 7th, 1778, and baptised at Westminster. In 1790 he was sent to Eton; he was, at the age of sixteen, given a commission in the 10th Hussars by his friend the Prince of Wales; after losing everything by gambling, he died at Caen, quite imbecile, in 1840.

BRUMMELL, WILLIAM, of Donnington Grove: born, 1776; Captain, 1803; resigned, 1805; married, 1800, to Miss Daniell, eldest daughter of James Daniell, of Wimpole Street, London. He purchased Donnington Grove after the death of James Pettit Andrews, who had built the house.

BRUMMELL, WILLIAM, of Wivenhoe House, Essex: died, 1853.

BULLEY, FRANCIS ARTHUR, of Reading: born, May 18th, 1808; son of John Bulley, surgeon, of Reading; brother of Dr. Bulley, of Magdalen College, Oxford. He practised for years in Reading, also brother of John Blagrave Bulley; Assistant Surgeon, November 20th, 1852; retired, 1855, when the Regiment went on active service to Corfu, though he did some deputy work for it for many years after; married, August 12th, 1840, to Louisa Nash (she died, 1893), they had four sons and five daughters ; he died, April 21st, 1883, age 74.

BULLEY, JOHN BLAGRAVE: born, September 25th, 1805 ; baptised, March 5th, 1806; Ensign, July 11th, 1825 ; Lieutenant, March 24th, 1828 ; Captain, May 16th, 1831 ; resigned, August, 1852 ; married, February 24th, 1824, to Mary Jervis Briscoe (she died, April 15th, 1856), they had no children ; he died, September 25th, 1864. His father married at St. Giles, Reading, April, 1802, "Mr. J. Bulley, surgeon, to Charlotte, daughter of the late Captain Pococke." "Died, 1826, at Upper George Street, London, Mary, relict of Mr. J. Bulley, of Reading, and daughter of the late Rear-Admiral Toll, of Fareham."

BUNNY, EDWARD JOHN, of Speen, and Slinfold, Sussex : born, 1828 ; son of Edward Brice Bunny, of Speen ; Lieut., November 1st, 1852; Captain in the Royal Sussex Light Infantry Militia in 1854; retired as Major and Hon. Colonel, 1883 ; married Mary St. John, only child of Robert Burnett Brander, and grand-daughter of Henry St. John. The Militia Register calls him "John Bunny." He took the name of St. John under his father-in-law's will, 1877.

BURGESS, BENJAMIN : Captain, F Troop, 1651.

BURNE, MALCOLM HILEY: third son of Newdigate Burne, of Alleway, Guildford, and grandson of the late Rev. and Right Hon. Viscount Sidmouth, and cousin to Colonel Blandy-Jenkins ; Second Lieutenant, May 16th, 1881 ; Captain, Royal Sussex Regiment, January 29th, 1884. Had a medal for Hazara (Black Mountain) Campaign. Has held Staff appointment of District Instructor of Signalling, Station Staff Officer, Regimental Paymaster, etc. He was killed by a fall of 600 feet down a Khud, May 10th, 1895, while shooting with a brother officer in the Native State of Chamba, in the Himalayas, and buried at Dharmsala, where a memorial has since been erected to his memory by his brother officers.

BURNE, KNIGHTLEY P.: born, November 10th, 1858; second son of Mr. and the Hon. Mrs. Newdigate Burne, of Albury, Guildford, and grandson of the late Rev. and Right Hon. Viscount Sidmouth; late Captain 72nd Regiment (Seaforth Highlanders); now of the Indian Staff Corps; medal and clasp and Khedive's Star for Egyptian Campaign (Tel-el-Kebir), 1883; medal and clasp for Upper Burmah, 1886, where he served with distinction, led the attack and carried the fortified stockade of Chanyone, the Camp at Octong, and other places; has served on the Staff as District Recruiting Officer for the Dogra Sikhs, etc., etc.; Lieutenant, Indian Civil Service, 1891; married, October 28th, 1887, Emma Marion, daughter of the late J. B. Summers, Esq., J.P., of Rose Moore, Pembrokeshire.

BURNINGHAM, JOHN: Captain, F Troop, 1651.

BURNETT, BENJAMIN: born, 1790; Ensign, February 17th, 1810; Lieutenant, April 5th, 1811; appointed to 3rd Dragoon Guards, January, 1812; married, 1812, Elizabeth Burnett.

BUTLER, ANDREW: Ensign, October 15th, 1806; displaced April 11th, 1807.

BUTLER, JOSEPH, of Kirby House, Newbury: Lieutenant, 1782; Captain, November 20th, 1786; resigned, October 19th, 1792; his eldest son, Captain Butler, of the Wilts Militia, married, 1812, at Guernsey, Eliza, only child· of Captain Dobsee, R.N.; died at Wantage, of apoplexy, 1823, in his 74th year.

BUTTER [or BUTLER], THOMAS WILLIAM, of Wokingham: Ensign, 1807.

BUTLER, —— : Major, 1655.

BYRNE, JOSEPH: Ensign, January 31st, 1800; Lieutenant, 1800.

CANE, ROBERT, of London: Ensign, 1797; Lieutenant, December 27th, 1796; resigned, 1798.

CANNON, —— : Captain.

CARDIFFE, WILLIAM: born. 1782; Ensign, March 18th, 1808; Lieutenant, October 6th, 1808; volunteered to 4th Foot, April 11th, 1809.

CAZENOVE, REGINALD FREDERICK: born, 1872; son of Frederick Cazenove, of Forest Grove, Bracknell, by ——— daughter of Colonel W. A. Orr, R.A., C.B., of Bridgeton, N.B.; educated at Eton; Second Lieutenant, January 24th, 1891; Lieutenant, May 9th, 1892; went to the Carabineers (6th Dragoon Guards), 1894, which he resigned almost immediately; married, 1895, Lilian, Dowager Duchess of Cromartie.

CAZENOVE, PERCY, of Warfield: born, 1875; second son of Frederick Cazenove, of Forest Grove, Bracknell; educated at Eton; Second Lieutenant, February 19th, 1894; Lieutenant, February 20th, 1895; resigned, 1896.

CERJAT, AUGUSTUS H. SIGISMUND DE: Ensign, March 10th, 1838; resigned, April 15th, 1852. The name de Cerjat is unusual in England; the following entry is from the register of St. George's, Hanover Square, "Married, 1800, Henry Andrew Cerjat, of Landough House, County Glamorgan, B., to Katherine Annabella Bristow." In 1802, Major Charles Cerjat became Lieutenant-Colonel of the 1st Dragoons.

CHAMBERLEYNE, A.: Second Lieutenant, March 23rd, 1897.

CHAPMAN, JOHN: born, 1788; Ensign, September 27th, 1809; Lieutenant, February 10th, 1810; volunteered to the Line. There were two John Chapmans in the Army, one a Lieutenant on half-pay, 56th Foot; the other, Captain in the 63rd (West Suffolk) Regiment. One of the Burgesses of Windsor, 1813, was a Mr. Chapman. Sir John Chapman was a surgeon in Windsor about that date. Probably connected with the Chapmans of South Hill, Delvin, County Westmeath.

CHAUVAL [or CHARWELL], EDWARD: born, 1790; Ensign, September 12th, 1812; appointed Ensign in 15th Foot. There was a Rev. A. R. Chauval, Rector of Great Stanmore, in 1824, Prebendary of St. Paul's.

CLANCY, RICHARD: born, 1777; Ensign, March 19th, 1807; volunteered to 9th Foot, September 3rd, 1807.

CLARKE, JOHN: born, 1785; Ensign, October 25th, 1808; Lieutenant, December 16th, 1808; volunteered to 4th Foot, April 12th, 1809. "Buried at Maidenhead, August 2nd, 1852, John Clarke, age 62." There was a William Clarke in the 4th Foot at Waterloo, gazetted Lieutenant in that Regiment July 28th, 1813.

CLASSON, HENRY: born, 1780; Ensign, August 2nd, 1803; resigned, 1804.

CLAVELAND [or CLEVELAND], WILLIAM, probably of Hare Hatch: Ensign, 1780.

CLAVER, JOSEPH: Captain, F Troop, 1651. If the F Troop meant the Faringdon Troop, then he may have come from that division of the County.

CLIMENSON, HENRY JOHN MONTAGUE, of Shiplake: born, December 15th, 1866; son of Rev. John Climenson, D.C.L., Vicar of Shiplake, by Emily Jane, only daughter of Hon. Spencer Dudley Montagu, thirteenth child of fourth Baron Rokeby, by Anne Louisa, only daughter of Sir Charles Flint, of the Irish Office, and widow of Joseph Jekyll, of Wargrave Hill; Lieutenant, April 14th, 1886; resigned his commission, May 14th, 1887; died at Lahad Datu, North Borneo, January 28th, 1891, where he was overseer on a tobacco estate.

COBHAM, ALEXANDER COBHAM, of Shinfield: born, 1808; Ensign, March 24th, 1828 (never joined); resigned, March 8th, 1831; married at Ealing, 1831, Jane Halse, second daughter of Richard Chambers, of Cradley Hall, County Hereford.

CODD, ROWLAND BENTINCK, formerly 17th Regiment: Adjutant, 1855. Afterwards Governor of Clerkenwell Prison.

COLEMAN, GEORGE THOMAS: Ensign, June 18th, 1832; nothing was heard of him in 1852.

COLES, WILLIAM: born, 1788; Ensign, February 15th, 1809; Lieutenant, February 10th, 1810; volunteered to the Line; Cornet, 25th Regiment Light Dragoons.

COLLYER, JOHN: born, 1787; Ensign, April 4th, 1810; resigned, April 6th, 1811; Lieutenant, 3rd Ceylon Regiment.

COLLIS, WILLIAM: probably eldest son of Rev. Samuel Collis, of Fort William; Ensign, February 25th, 1799. In 1809 Captain William Collis was promoted Major in the 27th Regiment. Captain and Adjutant, Royal Kerry Militia; married, 1814, Deborah, daughter of Dr. Crumpe, of Tralee; died, 1834.

COSTOBADIE, GERALD EDWARD, of Woolhampton: born, at Leicester, April 22nd, 1862; son of Major Costobadie, of Woolhampton Cottage; Second Lieutenant, February 21st, 1880; Lieutenant, July, 1880; went to the 2nd Battalion Royal North Lancashire Regiment, January 27th, 1883; Captain, January, 1894.

COX, FRANCIS RENELL, of Aldermaston: born 1827; son of Dr. Francis Cox, who, for twenty-seven years was surgeon at Aldermaston, and died in 1852; another son of Dr. Francis Cox, is Dr. Richard Cox, of Theale. Ensign, January 6th, 1855; Lieutenant, September 11th, 1855; Captain, April 16th, 1861; resigned, July 19th, 1872; married, 1853, Miss Augusta Jenkins. For many years they have lived at Boulogne, in France.

CRAVEN, EARL OF, WILLIAM: born, 1737; Colonel of the Berkshire Militia; Lord - Lieutenant and Custos Rotulorum of Berkshire, 1786; succeeded his uncle, 1769; married, 1767, Elizabeth, daughter of Earl of Berkeley (she married after his death the Margrave of Anspach and lived at Hampstead Marshall till her death). He died at Lausanne, in Switzerland, November 24th, 1791 ; age, 52.

CRAVEN, EARL OF, of Hampstead Marshall and Ashdown: Captain, February 14th, 1829 ; resigned, March 18th, 1831 ; died, 1833.

CRAVEN, EARL OF, WILLIAM : gazetted to 9th Battalion ; Garrison Major-General, 1805 ; A.D.C. to the King, 1798 ; Lord-Lieutenant of Berkshire, 1819-25 ; died, July 30th, 1825.

CRAVEN, HON. OSBERT WILLIAM, of Ashdown: born 1847 ; third son of William, Second Earl of Craven ; Lieutenant, May 16th, 1867 ; Captain, May 8th, 1878 ; resigned, June 18th, 1872 ; now commands Berkshire Yeomanry.

CRAVEN, HON. WILLIAM : Lieutenant, March 30th, 1787 ; Captain, 1787 ; Major, April 30th, 1797.

CRAVEN, HON. AUGUSTUS WILLIAM : born, May 3rd, 1858 ; eldest son of Hon. William George Craven, by Lady Mary Catherine Yorke, second daughter of Charles, Fourth Earl of Hardwick ; Lieutenant, August 28th, 1875 ; resigned, April 24th, 1880 ; married, 1880, Florence Champagne, daughter of General Corbet Cotton.

CRAVEN, HON. RUPERT CECIL : born, 1870 ; second son of George Grimston, Third Earl of Craven, by Evelyn Laura, second daughter of Viscount Barrington ; Second Lieutenant, January 7th, 1890 ; Lieutenant, February 7th, 1891 ; re-signed, March 27th, 1895 ; was in the Royal Navy before he joined the Militia.

CROFT, JAMES HENRY HERBERT : born, May 4th, 1840 ; second son of Archer James Croft, of Greenham, by his second wife, Elizabeth, daughter of Henry Boyle Deane ; Ensign, October 10th, 1857.

CROWE, DAVID, of Sindlesham: Ensign, 1795 ; Lieutenant, 1797 ; resigned, 1798.

CURTIS, THOMAS JOHN, of Abingdon : born, 1828 ; Ensign, November 30th, 1852 ; resigned, March 24th, 1855.

CURTIS, —— : Captain, 1651.

DALMER, FRANCIS: born, 1777; probably son of J. Dalmer, of Friar Street, Reading, whose daughter married in 1804; Ensign, April 25th, 1803; Lieutenant, July 21st, 1803; appointed to 23rd Foot, March 9th, 1804; Captain, 23rd Foot, December 10th, 1807; Major, August 26th, 1813, mentioned in the Waterloo Dispatches; fought at Waterloo as Lieut.-Colonel and attained the rank of Colonel; died, October 2nd, 1855.

DALZELL, ROBERT, of Tidmarsh: born, 1740; son of Gibson Dalzell, of Tidmarsh, whose father was General the Hon. Robert Dalzell, descended from the Earls of Carnwath. Robert Dalzell was heir to his grandfather, was educated at Westminster, and was a Gentleman Commoner at Christ Church, Oxford; married, December 1762, Jane, daughter of Colonel John Dodd. He was Patron of the living of St. Lawrence's Church, Reading. He lived at Toulouse, South of France. Died, 1821.

The present representative of the family is Miss Dalzell, of St. Alban's Priory, Wallingford.

DAMANT, GUYBON: born, 1788; Ensign, March 25th, 1812; Assistant Surgeon, March 26th, 1812; resigned, November 3rd, 1812.

DANVILL [or DARVALL,] CHARLES: Ensign, September 29th, 1807; appointed to the 10th Foot, January 28th, 1808. Probably related to Mr. Darvall, Solicitor of Reading.

DAVENPORT, JOHN INAM: born, 1787; Ensign, December, 9th, 1812; Lieutenant, April 30th, 1814; resigned, July 26th, 1832.

DAVIES, GEORGE: born, 1781; Ensign, July 10th, 1811; resigned, December 5th, 1811.

DAVIES, E., of Sandhurst: Second Lieutenant, August 15th, 1877; Lieutenant, August 20th, 1878; went to 81st Foot, April 17th, 1880.

DEANE, ARTHUR, of Waltham St. Lawrence: Ensign, May 3rd, 1831; Lieutenant, December 31st, 1852; resigned, October 31st, 1855.

DEANE, JOHN, of Ruscombe: Ensign, 1759; died, 1784, after a long illness, age 76.

DEANE, HENRY BOYLE, of Reading, afterwards of Hurst: son of Henry Deane, of Reading, by Lucy, daughter of John Wilder, of Nunhide, whose wife was Beaufoy Boyle; Ensign, February 22nd, 1793; Lieutenant, June 19th, 1793; Captain, February 14th, 1795; resigned, 1799; both he and his father afterwards took Commissions in the Woodley Volunteer Cavalry as privates.

DESBOROUGH, JOHN: Lieutenant, November 2nd, 1872 (never joined); resigned, November 20th, 1872.

DE VITRE, HENRY DENIS, of Charlton, near Wantage: born, 1831; son of M. T. Denis de Vitre, of Southwick Crescent, Hyde Park; Lieutenant, January 8th, 1863.

DODD, JOHN, of Swallowfield: only son of Randall Dodd, of Chester, by Margaret, daughter of William Glascour; Major, 1758; Lieut.-Colonel, 1762. He inherited a very large property from his great-aunt, Isabel, wife of Sir Samuel Dodd, Chief Baron of Exchequer. She was a daughter and co-heir of Sir Robert Croke, and her mother was one of the co-heiresses of Sir Peter Vanlore, of Tilehurst. The property was in several counties; that in Berks being in the parishes of Tilehurst, Tidmarsh, Beenham, Hampstead Norris, and West Compton. He came of age in 1737, and bought Swallowfield from Thomas Pitt for the sum of £20,770. The following year he married at Shinfield, Jane, daughter of Henry Le Coq St. Leger, of Trunkwell, and she died 1778. His second marriage was with Julia, daughter of Philip Jennings, of Plas Warren, County Salop. John Dodd was Member for Reading in various Parliaments, 1755-1780. He died February 11th, 1782, aged 65, at his house in Audley Square, after a severe

T

illness, with painful operations bravely borne with the calm fortitude and Christian resolution so eminently peculiar to him; buried at Swallowfield. A fine portrait of John Dodd is in the possession of General Parke.

DODD, ——: probably son of John of Swallowfield; Lieutenant, June 1789; resigned, 1789. John Dodd had only two sons; the eldest, John, was Colonel of the Guards, born about 1742; the younger, Harry, born about 1766, Captain of the 1st Life Guards, married Castellina, daughter of Warner Westerna, he died of consumption, October 29th, 1789, at Purley, in his twenty-fourth year.

DODWELL, J. [or T.] W.: born, 1792; probably son of the learned Dr. Dodwell of Shottesbrook (in 1785, died in The Close, Salisbury, age 75. Rev. Canon Dodwell, Archdeacon of Berks, Canon of Salisbury, Vicar of Bucklebury and White Waltham); Ensign, January 13th, 1813; died, December 5th.

DOE ——; Captain, 1651.

DOLMAN, THOMAS, of Shaw: baptised, 1657; second son of Sir Thomas Dolman, of Shaw, M.P. for Reading 1660, by Margery, daughter of John Hobilday, of Thornton, County Warwick; Knighted, 1703; married Dorothy, daughter of John Harrison, of Scarborough, relict of Henry Ball; died, 1711, without issue; described on his monument as "Collonoll of the Militia of this county."

DORAN, JAMES GODDARD: born, 1787 (this is, no doubt, the correct date); Captain, November 30th, 1815; resigned, April 13th, 1831. He founded G. S Navigation Company, and died, 1841, having sold Wellhouse a few years before his death.

DORAN, JAMES GODDARD, of Wellhouse; born 1783; Captain, February 11th, 1811; gazetted to the 14th Foot, December 26th, 1814. I believe, although the Register gives his age differently, after a year returned again to the Berkshire Militia. His father married in 1782, at St. James, London, Mrs. Doncastle, of Wellhouse, in Hampstead Norris.

She died, .1785, and he married, secondly, Miss Goddard; he died, 1792, age 47, leaving this son to inherit.

DOUGLAS, JAMES, of Hailingbury, County Hertford: born, 1825; son (or grandson) of James Douglas, M.D., of Reading; Ensign, November 30th, 1852; Lieutenant, January 5th, 1853; Captain, January 12th, 1855.

DOWNES, PERCY, of Newbury: nephew of Colonel Downes, of Donnington; Second Lieutenant, November 27th 1889; Lieutenant, February 7th, 1891; resigned, February 27th, 1895; married, at Folkestone, Miss Montgomery, daughter of Colonel Montgomery; went to Africa, 1895.

DUFFIELD CHARLES JOHN EDWARD, of Marcham Park: born, 1863; eldest son of Charles Philip Duffield, of Marcham Park, by Penelope, daughter of William Graham, of Fitzharris; Lieutenant, January 30th, 1886, resigned, February 18th, 1890.

DRUMMOND, ——, of Maidenhead: Second Lieutenant, resigned, 1896.

EAST, AUGUSTUS HENRY, of Woolley Hall, Maidenhead: born, August 24th, 1766; second son of Sir William East, of Hall Place, by Hannah, daughter of Henry Casamajor, of Tokington, County Gloucester; Captain, May 19th, 1798; Major, October 21st, 1801; resigned, October 16th, 1810; entered Trinity College, Oxford, 1787; married, December 29th, 1792, Caroline Anne, eldest daughter of George Vansittart, of Bisham Abbey. Major East met with a serious accident in 1803, while travelling in his curricle; he dropped the reins, and while striving to recover them he overbalanced himself and pitched on his head, and becoming entangled in the harness, was dragged some distance. He was much cut and bruised about the head, but was in a few days removed to London. Mr. Keate, the surgeon who attended him, pronouncing him out of danger. Died, 1828, at Bisham Abbey, which he rented.

T 2

EAST, GEORGE FREDERICK CLAYTON: born, October 18th, 1857; only son of the late Frederick Richard Clayton East, Captain 8th Madras Cavalry, who was fourth son of Sir East George Clayton East, of Hall Place; Lieutenant, June 28th, 1876; went to 25th Foot, August 14th, 1878; joined 2nd Battalion Royal Inniskilling Fusiliers.

EDWARDS, CHARLES MACKENZIE: from 1st Battalion; Adjutant-Captain, October 30th, 1890; Major, December 7th, 1891; returned to 1st Battalion, October, 1895. He was in the Siege of Kandahar, and gained a medal and one clasp, 1879-80; and another medal, with two clasps, in the Soudan Campaign, 1885-86.

EDWARDS, THOMAS HUGHES: Ensign, February 13th, 1799.

EGERTON, GEORGE ALGERNON: eldest son of Algernon Fulke Egerton, of Worster Old Hall, by Alice Louisa, daughter of Lord George Cavendish; Lieutenant, April 4th, 1883; resigned, March 6th, 1884.

ELLIOTT, THOMAS: born, 1772; Ensign, August 26th, 1801; Lieutenant, May 25th, 1803; volunteered to 52nd Foot, August 25th, 1807; married, Maria, daughter of Oliver Lloyd, of Coedmore.

ELLIOTT, GEORGE HENRY, of Binfield Park: born, 1789; son of Rev. George Henry Glasse, Rector of Hanwell (assumed the name of Elliott); Captain, 1813; Lieutenant, 20th Light Dragoons; Lieut.-Colonel of the Berks Militia; married, 1812, Mary Josephine, daughter of General Sir James Hay, commanding the Kent district.

ELWES, GEORGE, of Marcham: son of John Elwes, of Marcham, whose father, Robert Meggott, came to Marcham in 1717 (the name, Elwes, was assumed on succeeding to the property of Sir Harry Elwes, in Suffolk); Captain, 1779. George Elwes was known as the Berkshire miser; his father died when he was four years old, and he was over forty when he inherited his uncle's property. His son was in the Guards, probably it was the latter who was in the Berkshire Militia. Died, 1789.

EVELYN, ARTHUR: Captain, H Troop, August 23rd, 1650; Major, October; raised to Colonel, October 30th. He was Governor of Abingdon, 1646, then called Adjutant Evelin, and was entrusted to destroy the fortifications of Wallingford Castle, November 18th, 1652.

EVANS, HENRY: Lieutenant, 1779; Lieutenant in Captain William Sladden's Company, 1781; Adjutant; died, 1782.

EVERETT, FREDERICK, of Newbury: son of Rev. G. F. Everett, Rector of Shaw-cum-Donnington; born, 1839; Ensign, October 10th, 1857.

EYRE, MATTHEW: Colonel of a troop which was raised for the King, 1648.

EYRE, GEORGE BRAMSTON, of Welford. (See *Archer-Houblon*).

EYSTON, FRANCIS THOMAS, of Stanford Place: born, April 30th, 1853; eldest son of George Basil Eyston, of Stanford Place, near Faringdon, by Maria Theresa, third daughter of George Thomas Whitgreave, of Moseley Court, County Stafford; Lieutenant, February 17th, 1871; Captain, March 20th, 1876; resigned, April 6th, 1881; educated at St. Mary's College, Uscott, County Warwick; married, June 17th, 1880, Angela Vavasour, sister of the present Sir William Vavasour; they had no children; died, December 8th, 1888.

FENNELL, EDWARD: born, 1770; Ensign, 1800; Lieutenant, 1803; appointed in the Brunswick Fencibles, July 12th, 1803.

FENNELL, EDWARD: born, 1776; Adjutant, September 16th, 1806; resigned, February 22nd, 1811. He returned to the Militia after three years' service in the Fencibles. The dates of birth given show how untrustworthy the Regimental Officers' List is.

FINUCANE, MICHAEL: born, 1764; Ensign, 1804; displaced, 1805.

FLOYER, PETER, of Shinfield: Lieutenant, 1759; married the daughter of Sir James Clarke, of East Molesey (she inherited considerable property in Buckinghamshire on the death of her sister, Mrs. Rugge, 1768, and died in 1773); Peter Floyer died at Shinfield Place, 1778. "Peter Floyer, late Captain in Berkshire Militia, a gentleman not more admired for his affability to all classes of people than for the excellence of his heart and under-standing. He will be sincerely regretted by all who had the happiness of his acquaintance. His children lament a tender and affectionate parent: his companions, a true and faithful friend; and the world, a worthy and honest man."

FOLKESTONE, VISCOUNT: born, 1775; Captain, 1803; resigned, June 1st, 1805.

FOLKESTONE, VISCOUNT: born, 1782; Lieut.-Colonel, December 9th, 1812; resigned, 1817, when Colonel John Blagrave was replaced.

FOLKESTONE, VISCOUNT: born, 1841; Second Lieutenant, July 2nd, 1870; resigned, July 18th, 1872.

FONBLANQUE, JOHN: Ensign, 1780; married, at St. George's, Hanover Square, 1786, Miss Frances Caroline Fitzgerald, youngest daughter of Colonel Fitzgerald; their infant daughter died July 8th, 1804. He was a Barrister-at-Law, and afterwards Member for Camelford, 1802-6. John Fonblanque, who assumed the name of De Grenier before that of Fonblanque, 1828, was Senior King's Council, and Senior Bencher of the Middle Temple, having been called to the Bar, 1783. He published one or two works, and died in 1837. He was descended from an ancient family in Languedoc, and inherited the title of Marquis but never assumed it in England.

FORREST, THOMAS, of Binfield: born, 1782; Ensign, September 27th, 1803; Lieutenant, 1804; Captain, 1805;

resigned, 1810; married the daughter of Colonel Lowther, M.P.; "1808, — Forrest, of Binfield, to the eldest daughter of Colonel Lowther, M.P., Westmorland." Mrs. Forrest, widow of Admiral Forrest, of Binfield, died, June, 1802. Forrest Lodge, Bracknell, was sold to Sir Warwick Morshead. They are a Hampshire family.

FORREST, ARTHUR: Ensign, March 18th, 1831; Captain, December 23rd, 1831; resigned, April 7th, 1852.

FORREST, STANFORD: Ensign, August 5th, 1837.

FOWLER, ERNEST MORTIMER, of Brimpton: son of Captain George Fowler, R.N.; Second Lieutenant, February 11th, 1888; resigned, November 28th, 1890.

FRENCH, WILLIAM NATHANIEL: Ensign, 1779; Lieutenant, 1781; resigned, May 18th, 1790.

GARNETT, TAYLOR: born, 1766; Ensign, 1800; Lieutenant, 1802; resigned, 1803; married, 1795, " Dr. Garnett, of Harrogate, to Miss Cleveland, of Hare Hatch." Probably Regimental Surgeon or Chaplain. Canon Garnet was Canon of Windsor, 1793.

GILL, PHILIP (of Midgham?): born, 1758; Ensign, November 24th, 1779; Lieutenant, 1781; Captain, November 18th, 1786; resigned, 1798; Rector of Tidmarsh, which living was given him in 1785, by Robert Dalzell, Esq.; married, 1795, " Rev. Dr. Gill, Vicar of Rousham, to Miss Tounshend, sister to Edward Loveden, Esq."; died, 1825, in his 67th year. Probably was Regimental Chaplain. There were Gills of Basildon and Ashampstead.

GILL, JAMES: Ensign, 1780; Lieutenant, November 18th, 1786; resigned, 1798. No doubt a brother of Philip Gill.

GODDARD, VINCENT, of Reading: son of John Goddard, of Upham, Wilts, by Elizabeth, daughter of Sir John Fetiplace, of Besils Leigh (so says Byrne, in his account of " Berkshire Members of Parliament "), but I think he was probably grandson of John Goddard, and son of Richard Goddard, clothier, of Reading, christened at St. Mary's, 1612,

who died 1654, after which date Vincent Goddard continued
to rent the house his father had lived at, in Castle Street,
but he wanted an abatement of rent, which the Corporation
of the Town, of whom he rented the house, were reluctant
to grant; Captain, F Troop, August 23rd, 1650; Major,
October 30th, 1650. In 1612, Vincent, son of Richard
Goddard, was christened at St. Mary's, Reading, and,
according to the Corporation MSS., he had been appren-
ticed for seven years to his father. In 1656 he is styled
*Mr.* Vincent Goddard, so had evidently left the Militia.
He represented the County of Berks in Cromwell's Parliament, 1653.

GODDARD, RICHARD, of Reading: Captain, F Troop,
1651. The Goddards being Wilts may easily have belonged
to the Faringdon Troop. He was either a younger brother
or cousin of Vincent Goddard.

GOFFE, ——: Colonel; appointed, August, 1655, in
command of the Militia Forces of the Counties of Southampton, Sussex and Berkshire.

GOWER, JOHN LEVESON, of Bill Hill: Captain, May 4th,
1831; Major, July 12th, 1845; retired, September 20th,
1855; married, 1825, Charlotte Gertrude Elizabeth, second
daughter of Lady Harriett and the late Colonel Mitchell.
He was in the Navy.

GOWER, PHILIP LEVESON, of Bill Hill: Second Lieutenant, April 2nd, 1888; Lieutenant, February 7th, 1891;
went to the Sherwood Foresters, October 9th, 1891.

GOWER, C. C. LEVESON, of Bill Hill: Lieutenant, January
12th, 1884; went to the Royal Warwickshire Regiment,
April 28th, 1886.

GRAHAM, BIRCHALL GEORGE, of Newbury: born, November 11th, 1833; son of Robert Fuller Graham, solicitor
and town clerk of Newbury; Ensign, January 6th, 1855;
took one hundred volunteers to 33rd Foot (Duke of
Wellington's Regiment); was wounded in the Crimea;

served in Abyssinian expedition; died, a tea planter, at Darjeeling, Upper India, August, 1893. He was well known locally as an excellent cricketer.

GRAY, JOHN ROBIN, of Farley Hill: born, 1864; eldest son of Colonel Gray, of Farley Hill, and of Frankby, County Cheshire: Lieutenant, February 26th, 1884; went to Royal Irish Rifles, November 24th, 1885; returned as Captain, April 16th, 1889: married, 1891, Blanche, only daughter of the Rev. J. R. Fielden, Vicar of Honingham.

GREENE, RICHARD: Cornet, 1650.

GREENHEAD, CHARLES, 7, Bellevue Terrace, Haverford West: born, 1781; Surgeon, November 13th, 1813; resigned, February 24th, 1855.

GREENWAY, HENRY, of Trunkwell: Ensign, December 23rd, 1831; Captain, September 14th, 1837; resigned, January 20th, 1858. "Married, at Newbury, May 29th, 1802, Henry Greenway, of Henley-on-Thames, to Miss Woodroofe, eldest daughter of B. Woodroofe, of Newbury." Mrs. Greenway, relict of Henry Greenway, of Trunkwell, only died a couple of years ago.

GRENFELL, G. G.: born, 1786; Captain, February 1st, 1811; resigned, May 2nd, 1813.

GRENFELL, WILLIAM HENRY, of Taplow Court: born, 1855; Sub-Lieutenant, April 24th, 1874; never joined. He succeeded his grandfather to the Taplow Court property, 1867.

GRIBBLE, JOHN: born, 1772; Ensign, 1804; Lieutenant, 1804; resigned, 1805. He probably went to the Line.

GRIFFITH, CHRISTOPHER WILLIAM DARBY, of Padworth: born, 1860; only son of Christopher Darby Griffith, of Padworth; Second Lieutenant, May 8th, 1878; Lieutenant, June 25th, 1879; went to the Grenadier Guards, May 7th, 1880.

GRIFFIN, JOHN: born, 1782; Ensign, 1807; Lieutenant, 1808; resigned, 1808; married, 1887, Ethel Anne, only

daughter of Hon. Julian Fane. " Died, February, 1783, much
regretted by his friends and acquaintances, John Griffin,
of Wokingham, age 57, after a severe and prolonged
illness endured with fortitude." Probably related to Lord
Braybrooke.

GROVE, THOMAS : born, 1759; son of John Grove, of
Ferne, by his second wife, Phillipa, eldest daughter of Walter
Long, of Preshaw, County Hants; Lieutenant, 1779; Cap-
tain, 1781; resigned, February 9th, 1787; married, 1781,
Charlotte, daughter of Charles Pilford, of Effingham, County
Surrey; died, April 22nd, 1847. Probably related to
Thomas Grove, buried at St. Mary's, Reading, 1779.

GRUBB, GEORGE : born, 1789; Ensign, September 14th,
1811; Lieutenant, August 25th, 1812; resigned, April 29th,
1813. He was, I fancy, a younger brother of Lieutenant
Grubb of the Blues, who died of fever, in Spain, 1813, son
of W. H. Grubb, of Eastwell House, Wilts.

GUY, GEORGE : born, 1767; Ensign, January 19th, 1799;
Lieutenant, March 11th, 1799; Quarter-Master, March 19th,
1807. George, son of Rev. Mr. Guy, Vicar of Speen, gave
forty shillings per annum to Speen School, 1713. This was
probably a son of George Guy, of London. Probably went
to the Line. George Guy, aged 21, married, at Portsea,
1799, Jane Mortimore, daughter of John Grant; she was
a minor.

GUYENETT, FRANCIS : Lieutenant, 1798; resigned, May
18th, 1799.

HALLETT, WILLIAM, of Denford : born, 1784; Lieutenant,
August 26th, 1812; Captain, November 17th, 1812; resigned,
November 29th, 1813. Mr. Hallett bought Little Wittenham
Estate, 1787. William Hallett, of Denford Park, bought the
valuable property of Letcombe Basset, 1812; his wife died,
1833, in Southampton Street, Bloomsbury, after a happy
union of forty-eight years; he is described then as of Candy,
near Southampton.

HALLETT, W. (no doubt son of the above): Ensign, 1813; was living at Watchfield House in 1823.

HALLETT, GEORGE HUGHES, of Surbiton: born, 1832; Ensign, March 18th, 1855; afterwards Royal Artillery.

HAMILTON, ARTHUR: Ensign, 1805; Lieutenant, February 11th, 1806; appointed Quarter-Master 7th Garrison Battalion, April 21st, 1808. The Hon. Captain Hamilton sold Bear Place, 1780.

HAMMOND, RICHARD [or ROBERT]: fought in the first battle of Newbury, 1642, as Captain; Governor of Carisbrook Castle when King Charles was taken there a prisoner, 1647; Governor of Reading; High Steward of Reading. He married a daughter of John Hampden; died, 1656.

HANKEY, SIDNEY ALERS, of Heathlands, near Wokingham: born, 1847; Lieutenant, May 8th, 1871; resigned, May 16th, 1874; married, 1863, Louisa Fanny, daughter of Thomas Thornhill of Beddlesworth, County Norfolk; sold Heathlands, 1896, to Howard Palmer, of Reading.

HANKEY, SIDNEY THORNHILL, of Heathlands; born, 1869; eldest son of the above; Lieutenant, June 5th, 1886; went to the 2nd Life Guards, June 28th, 1890.

HANCE, JAMES: born, 1794; Ensign, April 29th, 1814; Lieutenant, October 15th, 1814; resigned, October, 1852.

HARGREAVES, ARTHUR, of Arborfield: born, August 20th, 1859; eldest son of Thomas Hargreaves, of Arborfield, by Sarah, daughter of Washington Jackson, Esq.; Lieutenant, May, 1877; died at Sunbury-on-Thames, October 25th, 1878.

HARGREAVES, ROBERT, of Maiden Erleigh: born, 1878; third son of John Hargreaves, of Maiden Erleigh and Whalley Abbey, by Jane, only daughter of Alexander Cobham-Cobham, of Shinfield; Lieutenant, February 7th, 1891; resigned, 1895.

HARRISON, J. S.; born, 1839; Ensign, January 24th, 1856; Lieutenant, February, 25th 1858.

HARVEY (OR HERVEY), LIONEL CHARLES: born, 1784; third son of Felton Lionel Hervey, by Selina Mary, heiress of Sir John Elwell, Bart.: educated at Eton; Ensign, 1805; Lieutenant, 1805; Captain, April 5th, 1805; Major, August 24th, 1826; married, 1825, Frances Mary, daughter of Vice-Admiral Thomas Wells; died, June 4th, 1843. His two elder brothers assumed the name of Bathurst, and were baronets.

HATCH, GEORGE, of New Windsor: Ensign, 1758; he was several times Mayor of Windsor, and died an old man in the autumn of 1800; one of the oldest members of the Corporation of Windsor.

HAY, ARTHUR WILLIAM HENRY, of Oakley Park, Eye, Suffolk: born, 1862; eldest son of Hon. Charles Rowley Hay, of Harewood Lodge, Sunninghill, by Arabella Augusta, daughter of Colonel W. H. Meyrick, grand-daughter of William Henry, first Duke of Cleveland; educated at Eton; Second Lieutenant, April 27th, 1880; Lieutenant, June 18th, 1881; Captain, May 22nd, 1886; Honorary Major, May 7th, 1895; served with the 2nd Battalion in Ireland and at Chatham in 1895; Major, December, 1896; married, November 24th, 1891, Mary, youngest daughter of Sir Edward Scott, fifth Baronet, of Lytchet Minster, County Dorset. In 1891 he succeeded to part of the Estates of his great-uncle, the Duke of Cleveland, and is now a Justice of the Peace for the Counties of Wilts and Suffolk, in which counties, and also in Gloucestershire, he owns property.

HAY, HARRY CLAUD FREDERICK, of Sunninghill: born, December 18th, 1864; brother of the above; Second Lieutenant, March 12th, 1894; Captain, April 19th, 1895; married, 1889, Lepel, youngest daughter of the late Captain Sayer, of the Royal Welsh Fusiliers.

HAYES, JOHN BEAUCHAMP, of Arborfield: born, 1848; only son of Rev. Sir John Warren Hayes, Bart., of Arborfield, by Ellen, daughter of George Beauchamp, of The Priory, Beech Hill; Lieutenant, April 18th, 1866; Captain, 12th Lancers; married, Julia, daughter of H. H. Hopkins; died, 1885, leaving three daughters.

HAWES, FRANCIS: born, 1746; Ensign, 1780; Lieutenant, November 17th, 1786; Captain, 1794; Captain-Lieutenant, November 24th, 1794; Quarter-Master, ——; displaced, February 28th, 1807. Probably the last owner of Purley Hall, which was sold to the Wilders in 1779, and some of the deeds mention Francis Hawes, linen draper, of Reading, and his mother, Mrs. Elizabeth Hawes. "In 1782, at his daughter's house in the Minories, died Francis Hawes, of Great Marlow, aged 86." "1775, died at the house of her mother in Charles Street, London, Miss Hawes, late of Purley Hall, aged 36."

HAWKINS, BENJAMIN: born, 1795; probably son of Benjamin Hawkins, Mayor of Newbury (Benjamin Hawkins died, 1825, aged 70); Ensign, September 9th, 1815; Lieutenant, September 28th, 1818; left about 1820.

HEATH, JAMES: born, 1789; Ensign, April 4th, 1812; resigned, December 14th, 1812.

HENDERSON, HAROLD GREENWOOD, of Buscot: born, October 29th, 1875; son of Alexander Henderson, of Buscot; educated at Eton; Second Lieutenant, March 3rd, 1894; Lieutenant, May 3rd, 1895; went to 1st Life Guards, 1897.

HERCEY, THOMAS FRANCIS JOHN LOVELACE, of Crutchfield: born, 1864; eldest son of Thomas Joseph Hercey, of Crutchfield, by Rowena Maria, daughter of William L. Pyne; Lieutenant, June 6th, 1883; resigned, April 27th, 1887.

HERCEY, FRANCIS HUGH GEORGE, of Crutchfield: younger brother of the above; Lieutenant, July 6th, 1886; went to the Royal West Surrey Regiment.

HILL, FREDERICK: Lieutenant, February 5th, 1799; resigned, 1800.

HILL, JOHN, of Barkham : Ensign, 1795 ; resigned, 1796.

HILL, SWANN, of London: Ensign, December, 1793; resigned, January, 1795.

HIPPESLEY, WILLIAM HENRY, of Sparsholt: born, 1855; second son of Henry Hippesley, of Lambourne, by his second wife, Elizabeth Mary, eldest daughter of Right Hon. Lawrence Sullivan ; Lieutenant, March 15th, 1872; resigned, April 5th, 1875; went to the Scots Greys; married Flora, fourth daughter of Mrs. Hargreaves, of Arborfield.

HODGE, BALDWIN : brother of Hermon - Hodge, of Wyfold; Second Lieutenant, January 29th, 1887; resigned, March 26th, 1887 (never joined).

HODGSON, W. S.: born, 1781; Ensign, November 28th, 1809; resigned, November 25th, 1810.

HODGSON, W. S.: born, 1783; Lieutenant, October 29th, 1811; resigned. June 14th, 1815. I cannot help thinking, in spite of the age being given differently, that these two Hodgsons are one and the same; the initials are alike. Perhaps, for some reason, he left the county, and returned after and re-entered as a Lieutenant.

HOLDSWORTH, FRANCIS ROBERT, of Dartmouth ; Lieutenant, March 12th, 1797; Captain, October 18th, 1798; went to the 15th Foot, 1799, with his company.

HOLDEN, HARRY WINTON: from 1st Battalion; Adjutant, October 1st, 1885; rejoined 2nd Battalion, October 30th, 1890; married, at Gibraltar, 1884; died, 1893.

HOLLAND, J. S.: born, 1779; Ensign, September 25th, 1809; Lieutenant, January 10th, 1810; displaced, February 28th, 1811. Sir Nathaniel Holland purchased the Manor of Little Wittenham early in this century.

HOLLAND, THOMAS: born, 1777; Ensign, March 28th, 1803; resigned, April 27th ; probably went to the Line. Thomas E. H. Holland was Ensign in 4th Foot, gazetted December 9th, 1813.

HOLLAND, EARL OF, BARON KENSINGTON, HENRY RICH, High Steward of Reading (three who held the office perished on the scaffold); second son of the first Earl of Warwick; Lord-Lieutenant for Berkshire, 1632; Constable of Windsor; Knight of St. George; Groom of the Stole; Justice in Eyre; married Isabel, daughter and heiress of the Copes, from whom came the Manor of Kensington. He fought first on one side, then on the other, in the Civil War, and consequently was not appreciated by either party. His portrait was painted by Vandycke. He was said to be the handsomest man at Court. Executed for High Treason, March 9th, 1648.

HOLLOWAY, BENJAMIN, JUN., of Charlbury, County Oxon : Ensign, 1795; Lieutenant, December 19th, 1795; resigned, June, 1798; Benjamin Holloway, of Lee Place, married Susanna, daughter of Richard Wykham, of Swalcliffe, by his wife, Alicia, daughter of Rev. Richard Fiennes.

HOLT, GEORGE : born, 1778; probably son of John Holt, of Westmorland and Enfield, and Tottenham, Middlesex, descended from Sir Thomas Holt, who married (time of Charles I.) the daughter of John Peacock, of Cumnor; Ensign, January 28th, 1808; resigned, June 14th, 1808; probably went to the Line.

HOMFRAY, JOHN GLYNNE RICHARDS, of Penlynn Castle, County Glamorgan; born, 1861; eldest son of John Richards Homfray, of Penlynn Castle (who died 1882), by Mary Elizabeth, daughter of Sir Glynne Earle Welby-Gregory, Bart.; Second Lieutenant, April 19th, 1880; Lieutenant, January 12th, 1881.

HOMFRAY, H. R.: brother of the above; Lieutenant, March 7th, 1884; went to Royal Irish Rifles, November 24th, 1885.

HOPKINS, EDMUND JOHN ROBERT, of Tidmarsh : born, April 8th, 1856; eldest son of Robert John Hopkins, of Tidmarsh, by Elizabeth Clara, daughter of Rev. David

Rodney Murray; Lieutenant, May 12th, 1875; resigned, July 18th, 1877; educated at Eton and St. John's College, Oxford; died, June 2nd, 1861.

HORNE, ARTHUR: Ensign, F Troop, August 23rd, 1650; probably the Militia troop of Faringdon.

HORWARD, J. J.: born, 1839; Ensign, June 8th, 1859.

HOUBLON. (See *Archer*).

HOUGHTON, THOMAS ALDERMAN, of Broom Hall, Sunninghill; born, 1826; only child of John Houghton, of Armsworth, Hants, by Anne Sophia, eldest daughter of Stephen Shelden; Ensign, August 8th, 1853; resigned, January 16th, 1855; married Mary Cecilia Wakefield, daughter of Richard Attree, of Blackmore, Hants.

HOWARD, THOMAS AUBREY, of Yattendon: born, November 13th, 1802; eldest son of Thomas Howard, of Yattendon House, by a daughter of — Aubrey, Esq.; Captain, December 31st, 1852; resigned, January 12th, 1855; married Charlotte Mary, daughter of — Corrance, from Leicestershire, who died 1864. Mr. Howard died at Newbury, 1882.

HUNTER, BART., SIR CLAUDIUS STEPHEN PAUL, of Mortimer: born, 1825, at Ghazepore, East Indies; son of John Hunter, E.I.S., who was son of Sir Claudius Stephen Hunter, of Mortimer; Captain, October 8th, 1852; resigned, 1856; afterwards Lieut.-Colonel of the Berkshire Volunteers; married, 1855, Constance, daughter of William George Ives Bosanquet. He died January 7th, 1890.

HUNTER, BART., SIR CHARLES RODERICK, of Mortimer: born, 1858; eldest son of the above; Sub-Lieutenant, September 22nd, 1875 (never joined); went to the Rifle Brigade; married, 1887, Agnes Lillie, eldest daughter of Adam Kennard, of Crawley Court, Hants.

IMHOFF, CHARLES, of London: Lieutenant, January 24th, 1794; Captain, January 6th, 1796; Major, July, 1797;

resigned, February 4th, 1799. "February 14th, 1853, at Darlesford House, Worcestershire, aged 86, General Sir Charles Imhoff, Knight of St. Joachim." Sir Charles Imhoff, though of German extraction, was, we believe, a native of this country, and related to the celebrated Warren Hastings, who was a native of Darlesford. Perhaps he was son of General Imhoff, whose regiment was one of those who received the Princess Augusta, daughter of George III., in 1764, when as the bride of the Prince of Brunswick she arrived at that town. In 1786, he was recommended by Queen Charlotte to the notice of the reigning Prince of Waldeck, and was appointed by His Serene Highness to the command of a company in one of his regiments, which he joined, in 1787, at Arolsen, the capital of Waldeck. He remained in Germany for some years; but, having completed his military education, returned to England at the commencement of the war in 1793, and accepted a commission in the Berkshire Militia, which he quitted a Captain in 1798, and then purchased a troop in the 1st Regiment of Life Guards, by commission dated April 4th, 1799. In 1801, he became Major in the 4th Foot; and, on February 5th, 1802, Lieut.-Colonel in the same regiment. At the Peace of 1802, he again visited the Prince of Waldeck, but returned home from Berlin at the renewal of the war. He continued on half-pay until 1807, when he was, for a short time, Inspecting Field Officer of the Volunteers of the North Inland District of Nottingham; and was next appointed Lieut.-Colonel of the 4th Garrison Battalion, stationed at Jersey. He retained that command in the Channel Islands until June, 1812, when he was placed on the Staff as Inspecting Field Officer of the Guernsey Militia; and, after having occasionally officiated as Commanding Officer of the Garrison during the absence of the Lieut.-Governor, he was regularly sworn into that office on June 25th, 1814, and exercised its functions until

U

August 20th following, He was successively promoted to
the rank of Colonel in the Army, in 1818; Major-General,
1814; Lieut.-General, 1830; and General, 1846. On May
18th, 1807, he received the Royal Licence to accept the
insignia of a Grand Commander of the Order of St.
Joachim, and from that period he had enjoyed the titular
distinction of a Knight in this country—the regulation to
the contrary, with respect to Foreign Orders of Knighthood,
not being issued until the year 1813. He was one of the
Stewards for Westminster School 1828. Sir Charles Imhoff
married, February 19th, 1795, Charlotte, sixth daughter of
Sir Charles William Blunt, Bart.; she died, March 14th,
1847. Warren Hastings rented Purley Hall during his
celebrated trial; Charles Imhoff was said to be his stepson.
A portrait of a Colonel Blunt is at Purley; said to have
been a connection of Mrs. John Wilder, whose husband
was Major in the Berkshire Militia.

INCE, HENRY [James given in the Army List as his
Christian name]: born, 1795; Ensign, December 21st, 1815;
Lieutenant, July 2nd, 1821; resigned, October, 1852. His
name is omitted in the Army List after 1825.

INCE, HENRY ROBERT, of Westminster: Ensign, March
15th, 1793; left before 1797.

ISHERWOOD, RICHARD: born, 1784; Captain, June 26th,
1813; dismissed, September 15th, 1815. Probably son of
Henry Isherwood, M.P. for New Windsor, 1796, who died
the following year.

JENKINS, JOHN BLANDY, of Kingston and Llanharran:
born, 1839; Ensign, October, 1857; Lieutenant, April 16th,
1864; Captain, April 18th, 1866; Major, March 29th, 1873;
Lieut.-Colonel, July 18th, 1885; Honorary Colonel, January
29th, 1887; resigned, June 16th, 1888; twice married
firstly, Martha Alice, daughter of Charles Wilson Faber;
secondly, February 13th, 1897, at St. Gappan's, Glanmire,
County Cork, Elizabeth Norah Drury, fourth daughter of

the late Major-General George Drury, R.M.L.I., of North Huish, County Devon.

JENKINS, JOHN BLANDY, of Kingston: born, 1864; Lieutenant, November 18th, 1882; resigned, November 20th, 1887; married, 1888, Helen, only daughter of Thomas Duffield.

JOHNSTONE, CHARLES JOHN: born, 1877; second son of George Charles Keppel Johnstone, Lieut.-Colonel Grenadier Guards, by Agnes Caroline, daughter of Thomas Chamberlayne, of Cranbury Park and Weston Grove, Hants; Second Lieutenant, 1896; Lieutenant, 1897.

JONES, JOHN: born, 1773; Ensign, September 30th, 1803; Lieutenant, 1803; displaced, March 7th, 1804; married, 1805, "Lieutenant John Jones, of 36th Foot, to Miss Pye, daughter of Henry James Pye."

JUDD, STEPHEN: born, 1772; Ensign, 1803; Lieutenant, November 26th, 1805; Assistant Surgeon, July 25th, 1803; resigned, 1852.

JUSTICE, THOMAS, of Sutton Courtney: Lieutenant, 1758; married, November 24th, 1763, Catherine, daughter of Thomas Goodlake, of Letcombe Regis; "Died, June 1777, at his house at Appleford, near Abingdon, Thomas Justice, Esq." Another Thomas Justice died at Appleford, 1789.

JUSTICE, .THOMAS, of Sutton Courtney: Captain; died at Sutton Courtney, December, 1802, age 71.

KEARNEY, HENRY JOHN, of White Waltham: Colonel of Beynhurst Volunteers, 1803; Lieut.-Colonel of the 2nd Battalion of Militia, otherwise called the "Local Militia," 1809, it consisted of eight companies. "Died, July 20th, 1827, at White Waltham, in his 80th year, Henry John Kearney, Lieut.-Colonel of the 2nd Berkshire Militia. He was great-nephew to the first, and son-in-law to the second Duke of Chandos. He was son of the Rev. John Kearney, D.D., by Henrietta, fifth daughter of the Hon. and Rev.

Henry Bridges, Archdeacon of Rochester. Married : firstly, in 1778, his second cousin, Lady Augusta Brydges, daughter of Henry, second Duke of Chandos; she died childless a little more than a year after her marriage; and he married, secondly, the daughter and heiress of Joseph Banks, of Lincolns Inn, Chancellor of York. The Colonel embraced the Military profession at an early age, and served under General Elliott at the siege of Gibraltar."

KEEPE, ANDREW : Lieutenant, F. Troop, 1650.

KENRICK, WILLIAM, of Tilehurst: Captain, 1667. The Kenricks owned the property of Tilehurst. The daughter and heiress of Sir William Kenrick (who married, 1679, Grace, heiress of Peter Kibblewhite, of Swindon,) last of the family married Benjamin Child. She was the " Berkshire Lady," of whom the romantic story is told, that she fell in love with the handsome Reading attorney at a wedding and sent him a challenge. At the appointed place of meeting, she arrived masked, and gave him the choice of fighting the duel or marrying her; he chose the latter, wisely, and she proved so devoted a wife that he was broken-hearted after her death, and sold Calcot to John Blagrave. The story of the " Berkshire Lady " is prettily told in a novel written by Katherine Macquoid.

KEY, THOMAS : Ensign, December 18th, 1798 : probably son or grandson of Rev. Thomas Key, Canon of Windsor and Vicar of Upton Church, Bucks, who died, 1760. Thomas Keays, of Speen, married, at St. George's, Hanover Square, 1797, to Sarah Davis.

KINGE, JOHN : Muster-master of Reading, March, 1626.

KING, AUGUSTUS HENRY W., of Warfield; born, 1865 ; son of Captain William Wallis King, 12th Lancers; by Katherine, daughter of John Stuart Sullivan ; Lieutenant, January 26th, 1884; Captain, February 15th, 1888 ; resigned, February 19th, 1889.

KINNERSLEY, WILLIAM THOMAS, of Binfield: Lieutenant, February 24th, 1863; Captain, May 4th, 1871; resigned, March 24th, 1872; married Rose Bertha, eldest daughter of M. A. Bazille Corbin; died, 1876.

KNOLLYS, ——, of Stanford-in-the-Vale: Captain in the Abingdon Division of the Trained Bands. He was brother to the Earl of Banbury. Died, 1640.

KNOX, ARTHUR, of Sonning: Second Lieutenant, January 19th, 1878; Lieutenant, June 25th, 1879; resigned, January 18th, 1882.

KNIPE, CHRISTOPHER, of Frilsham: born, 1871; son of General Knipe, who was renting a house at Frilsham, which he left in 1894; Second Lieutenant, March 28th, 1891; Lieutenant, November 25th, 1893; resigned, February 3rd, 1894.

LANE, JAMES HENRY: born, 1835; Ensign, November 15th, 1855.

LANG, FREDERICK HENRY, from 34th Regiment: Adjutant, July 17th, 1858.

LANGFORD, JOSEPH: Ensign, 1762.

LEE, BART., SIR GEORGE PHILIP, of Windlesham: Ensign, July, 1840; married, 1843, Charlotte, daughter of John Ede; died, 1870.

LEE, HENRY PINCKE, of Woolley Firs, White Waltham: eldest son ⸱of John Lee, of Woolley Firs, by Dorothy, daughter and heiress of Thomas Hasker, of Kempshott, County Hants, baptised at White Waltham, September 4th, 1770; niece and heiress of Henry Pincke, of Arborfield; Ensign, 1793; Lieutenant, August, 1793; resigned, 1795; married Matilda, daughter of Stanlake Batson, of Winkfield Place; buried at·White Waltham, March 8th, 1826; age, 55. His eldest surviving son is Rev. Stanlake, Rector of Broughton, County Hants.

LEYCESTER, HENRY HANMER, of Cookham: born, 1808; second son of George Hanmer Leycester, of White Place,

Cookham, by Charlotte Jemima, daughter of Hans Winthrop Mortimer, of Caldwell, County Derby; Lieutenant, January 1853; Captain, October 10th, 1857; died, January 22nd, 1862. He was so tall, that he was called "Long Leycester," and when he went to Corfu the berth had to be cut to fit him.

LEYCESTER, OSWALD WALDEN: fourth son of George Hanmer Leycester, of White Place, by Charlotte Jemima, ·daughter of Hans Winthrop Mortimer, of Caldwell, County Derby; Ensign, May 4th, 1831; resigned, June 7th, 1833.

LLOYD, WILLIAM, of Shrewsbury: Lieutenant, 1795.

LOVEDEN, EDWARD LOVEDEN, of Buscot: Captain, 1779; Lieut.-Colonel, 1794. He married three times. By his first wife, married in 1773, Margaret, daughter and heiress of Lewis Pryse, of Woodstock, County Oxon, and Gogerthan, he had a son, Pryse Loveden. Colonel Loveden was M.P, for Abingdon. A lawsuit was brought against him in 1799, by Mrs. Elizabeth Cotterell, inn-keeper of Pangbourne, for election expenses. Her bill was for £157, of which he paid £105 into court; and the jury decided against her that the charges were absurd. Edward Loveden Loveden died, January 4th, 1822, at Buscot Park, in his 72nd year. He was L.D. of the University of Oxford, F.R.S. and F.S.A., and likewise member of several other scientific societies. He had, until the last few years of his life, enjoyed almost un-interrupted good health. His miniature has been engraved; a copy of it was kindly sent me by his descendant, Sir Pryse Pryse, of Gogerthan.

LOVEDEN, PRYSE, of Woodstock and Buscot: son of the above; Ensign, 1794. He married twice: his first wife was Harriet, daughter of William, second Viscount Ashbrook, she died childless, 1813; by his second wife, Jane, daughter of Peter Cavallier, of Gisborough, he had a son born, 1815.

Pryse Loveden took the name of his ancestors, and thus became Pryse Pryse, of Gogerthan. He died, 1849.

LOVELACE, BARON JOHN, of Hurley: Lord-Lieutenant of Berkshire; died at Woodstock, September 24th, 1670; buried at Hurley.

LUSH, JOHN: Captain in F Troop, 1650. Perhaps Faringdon.

LYNCH, EDWARD: Captain from 13th Light Infantry; Adjutant, February 8th, 1855.

MACPHERSON, LACHLAN: Ensign, December 17th, 1798. Perhaps related to the Neale Vansittarts.

MADOCKS, JOHN EDWARD, of Vron Iw., County Denbigh (described as of Mount Mased, Kent): Ensign, October 19th, 1792; resigned, 1793. He married twice; his second wife was Elizabeth, daughter of William, eleventh Earl of Craven. She died in 1779. He was M.P. for Denbigh. His eldest son was also John Edward Madocks. He married, 1817, Sidney, daughter of Abraham Robart Robartes, of London. She died in 1852, her husband having died in 1837.

MAGRATH, PHILIP: Ensign, 1800.

MAITLAND, THOMAS FULLER, of Wargrave and Garth, County Radnor; born, 1818; third son of Ebenezer Fuller Maitland, of Stanstead, Essex. and Park Place, Berkshire, by Bertha, grand-daughter and heiress of William Fuller, of Ponders End; Lieutenant, January 11th, 1853; Captain, September 25th, 1857; Major, May 8th, 1865; resigned, April 12th, 1871; married, 1842, Anna, only daughter of Captain A. B. Valpy, R.N., of Blagdon, County Somerset.

MARTIN [or MARTEN], HENRY: Colonel. A Commonwealth officer, one of the Regicides. Probably of Berkshire descent, as in 1614 Henry Martin had property at Clewer and Bray.

MARTIN-ATKINS, ATKINS EDWIN, of Kingston Lisle; born, 1778; eldest son of Edwin Martin-Atkins, of Kingston

Lisle, by Ellen Frances Halhed ; Captain, June 5th, 1799 ; married, 1806, Anne, daughter of Major Cook ; died, May 1st, 1825.

MARTIN-ATKINS, WILLIAM HASTINGS, of Farley Castle: born, 1810; was christened after his godfather, Warren Hastings; second son of Atkins Edwin Martin-Atkins, of Kingston Lisle, by Anne, second daughter of Major Cook, descended from David Martin, a French Protestant Divine, born 1639, who, on the Revocation of the Edict of Nantes, took refuge in Utrecht, becoming Pastor of the Walloon Church, and died in 1721 ; Captain, April 15th, 1861 ; resigned, May 7th, 1871. Married first, 1844, Diana Mary, second daughter of the Rev. James Wyld, and widow of John Tyrell, of Hew, who died 1862 ; and second, 1865, Georgiana, widow of Edward Lloyd Edwards, of Cerrig Llwydion, and eldest daughter of George E. Beauchamp, of the Priory, Beech Hill, and Thetford, County Norfolk, who died 1881. The name, Atkins, was assumed by Letters Patent, 1792, by Edwin Martin, the great-grandson of David Martin.

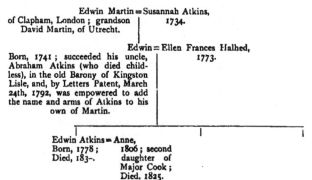

MARTIN-ATKINS, FRANCIS: born, 1852 ; Lieutenant, May 4th, 1871 ; Captain, June 9th, 1877 ; married Miss Johnstone.

MATON, CHARLES JOSEPH: resigned, 1795. "Married, 1783, at St. George's, Hanover Square, Mr. Maton, of Hartley Row, Hants, to Miss Anne Head, of Newbury, with a fortune of £30,000." "Married at St. George's, Hanover Square, 1790, Joseph Maton to Miss Ann Pinnell." These are the only entries I find of the name Maton.

MAURICE, DAVID BLAKE, of Reading: eldest son of Dr. Oliver Maurice, of London Street; Lieutenant, June 19th, 1885; went to 1st Battalion, of which he is now Adjutant.

MAYES, WILLIAM: born, 1787; Ensign, June 20th, 1809; Lieutenant, August 23rd, 1809; Volunteered to 10th Foot, May 1st, 1811.

MEARS, ELLIS, of Southampton; Ensign, February 15th, 1795; Lieutenant, 1796; resigned, February, 1796.

MEDLEY, A. L., of Faringdon: born, 1839; Ensign, November 14th, 1857.

METER, JOSEPH CHARLES, of New Sarum: Ensign, December, 1793.

MEYRICK, SAMUEL: born, 1783; probably he was son of Rev. Edward Meyrick, Master of Hungerford Grammar School, who married, 1777, at St. George's, Hanover Square, Miss Greaves, of Great Maddox Street, London; unless it was Sir Samuel Rush Meyrick, Bart., of Goodrich Court; Ensign, 1804; Lieutenant, 1805; resigned, May 7th, 1805; his age is given in the Militia Register as 28, but it is not reliable and he may have been younger. Sir John Rush lived at Streatley the middle of 18th century.

MICKLIN [or MICKLEM], ROBERT, of Hurley, afterwards of Stratford, near Salisbury: born, 1785; perhaps son of Robert Micklem, Mayor of Reading, who died in his mayoralty, 1793; Lieutenant, June 26th, 1812; resigned, December 26th, 1812; married at St. James' Church, 1815, Miss Cruthwell, only daughter of the late Mr. Richard Cruthwell, printer of the *Bath Chronicle*; she died, 1818, at Sonning, only 27 years of age.

MILNE, JOHN : Quarter-Master Sergeant in 72nd Highlanders, then in the Berks Militia ; promoted Quarter-Master, June, 1855.

MILMAN, WILLIAM GEORGE, of Levaton Woodland, County Devon: born, 1781; eldest son of Sir Francis Milman, by Frances, daughter and heiress of William Hart, of Stapleton, County Gloucester; Captain, 1808; resigned, 1812; married, 1809, Elizabeth Hurrey, only daughter of Robert Alderson, Recorder of Ipswich; succeeded his father as 2nd Baronet, 1821; died, August 21st, 1857.

MONCK, WILLIAM BERKELEY, of Coley: born, 1842; eldest son of John Bligh Monck, of Coley, by Elizabeth, daughter of Rev. Wildman Yates, Vicar of St. Mary's, Reading; Lieutenant, February 22nd, 1863; Captain, May 20th, 1870; resigned, April 26th, 1873; married, 1872, Althea Paulina Louisa, eldest daughter of Charles Alexander Fanshawe, Esq.

MORLAND, WALTER HOLROYD: born, October 31st, 1847; fifth son of George Bowes Morland, of Abingdon, Clerk of Peace, by Marie, daughter of J. Thornhill, Woodleys, County Oxon; Lieutenant, August 25th, 1870; Captain, February 5th, 1873; resigned, February 1st, 1879; died, July 20th, 1880.

MORLAND, GEORGE WILLIAM, of Abingdon: born, August 8th, 1839; second son of G. B. Morland and Maria (Thornhill,) as above; Ensign, September 16th, 1855; obtained a commission in 6th Regiment in 1857; died at Amritgar, India, September 23rd, 1874.

MORETON, JOHN : born, 1776; probably son or grandson of John Moreton, M.P. for Abingdon, 1754-62, who was made Chief Justice of Chester, 1762; his widow died in 1803; Ensign, July 2nd, 1803; resigned, October 7th, 1803. In 1817, a Thomas Moreton lived near Pangbourne, from his house the unfortunate Dr. Lonquet was murdered.

MORRES, E. J., of Wokingham ; born, 1830 ; son of Rev. T. Morres, of Wokingham ; Ensign, January 16th, 1855 ; resigned, May 17th, 1855.

MORRICE, CHARLES : Ensign, December 29th, 1792. In 1774, a marriage was celebrated between Rev. Dr. Morrice and Miss Hatch, of Windsor. She was probably the daughter of George Hatch, of New Windsor, who was Ensign in the Militia, 1758. Rev. Charles Morrice was Chaplain of Windsor.; Rev. Charles Morrice was appointed Chaplain of the Berks Fencible Cavalry in 1794. Colonel Charles Morrice, of 69th South Lincolnshire Regiment, was killed at Quatrebras.

MORSHEAD, BART., SIR WARWICK CHARLES, of Forest Lodge, Bracknell : born, 1824 ; only child of Sir Frederick J. Morshead, of Trenant, Cornwall, by Jane, daughter of Robert Warwick, of Warwick Hall, Cumberland ; Major, August 10th, 1863 ; resigned, March 4th, 1864 (never joined) ; married, 1854, Selina, daughter of Rev. William Vernon Harcourt, of Nuneham, she died, 1883 ; secondly, 1887, Sarah Elizabeth, second daughter of Montague Wilmott, sister to Sir Robert Wilmott, Bart.

MOORE, CHARLES : born, 1779 ; Ensign, June 7th, 1803 ; resigned, June 27th, 1803. In 1806, Charles Moore was auditor of Public Accounts.

MOWBRAY, ROBERT GRAY CORNISH, of Mortimer : born in London, May 21st, 1850 ; eldest son of Sir John Cornish Mowbray, Bart., of Warennes Wood, Mortimer, by Elizabeth Gray, only child of George Isaac Mowbray, of Bishop-wearmouth, whose name he assumed by Royal License Lieutenant, June 1st, 1872 ; resigned, April 29th, 1874 (never joined) ; educated at Eton, 1863-8 and Baliol College, Oxford, 1868-72 ; Fellow of All Souls, Oxford, 1873 ; Barrister-at-Law, Inner Temple, 1876 ; Member for Prestwich Division, S. E. Lancashire, 1886-95 ; Secretary to the Royal Commission on the Stock Exchange, 1876 ; Private Secretary to Mr. Goschen, Chancellor of the Exchequer, 1888-92 ; Member of Royal Commission on Opium, 1894 ; also Royal Commission on the Finances of India, 1896.

MURPHY, T. C., Captain, 66th Regiment; Adjutant, September 20th, 1875; rejoined 2nd Battalion, September, 1880; retired with the rank of Colonel. He afterwards lived in Reading; removed to Southsea, 1895.

NEPEAN, CHARLES EVAN MOLYNEUX YORKE: born, 1867 ; only son of Rev. Sir Evan Nepean, Bart., sometime Rector of Appleshaw, Hants, now residing at Bourne-mouth, by Maria Theresa, second daughter of Rev. F. Morgan-Payler, Rector of Willey ; Second Lieutenant, February 19th, 1887 ; Lieutenant, November 17th, 1888 ; Captain, February 18th, 1891 ; married, November, 1896, at Heytesbury, Mary Winifred, only daughter of Rev. W. J. Swayne, Vicar of Heytesbury.

NEWBOLT, JOHN THOMAS, of Wokingham : born, 1765 ; son of William Newbolt, by Anne Kent, of Wokingham (who died, 1786) ; Ensign, 1792 ; Lieutenant, March 15th, 1793 ; resigned, April 4th, 1804 ; Knighted by William IV for services rendered to Leopold I, (afterwards King of the Belgians, in 1831) ; married Catherine Dennis (who died, 1811, age 40) ; secondly, Miss Baldwin. They lived in London till 1802, Wokingham 1813, then went to Ghent and Brussels. Their daughter, Maria Newbolt, lived with the Duke and Duchess of Maxilien, of Bavaria, as governess to their daughter, the present Empress of Austria ; she married later the Count de Spietti. He had besides the two sons who were in the Berks Militia, Lieutenant Charles Kent Newbolt, R.N. (who brought home the Frigate *Menelaus*, with the body of Sir Peter Parker), and Colonel George Newbolt, served in India 37 years in the 31st Native Infantry and had medals for Ghuznee and Chillian-wallah. He died, 1837, at Brussels ; leaving six sons and four daughters.

NEWBOLT, WILLIAM KENT, M.D., of Barnstaple: born, 1786; son of the above; Ensign, January 30th, 1809; Lieutenant, June 20th, 1809; Surgeon's Mate; resigned, February 23rd, 1812; married Elizabeth Olivia Morrison. Mr. Henry Newbolt gives his wife's name as Louisa Maria Hyde. His commission, signed by Lord Radnor, is in the possession of his grandson, George P. Newbolt, F.R.C.S., of 42, Catherine Street, Liverpool (who is also a Physician). After leaving the Berkshire Militia, William Newbolt studied anatomy, &c., in London, under Sir Joshua Brookes, and later on joined the North Devon Militia. In 1818, he acted as second to Lieutenant O'Callaghan in the duel with Lieutenant O'Brien, in which the latter was killed. The seconds were imprisoned for some months as a punishment. After the North Devon Militia was disbanded in 1820, he practised as a medical man and died at Bath, 1859. His name is mentioned in the *Gentleman's Magazine* as having served in the Peninsular War.

NEWBOLT, FRANK N.: born, 1791; son of John Thomas Newbolt; Ensign, November 26th, 1811; Lieutenant, January 13th, 1813; appointed to the Royal Waggon Train, December 25th, 1813; married Jane Douglas; died, 1884, leaving a son and daughter.

NEWBURY [or NEWBERY], JOHN, of Heathfield Park, Sussex: Lieutenant, December 19th, 1795. In 1806, John Newbury, Lieut.-Colonel of the Sussex Militia, married at Rothwell, near Leeds, Miss Cleaver, daughter of Rev. Dr. Cleaver, of Malton, County York.

NEWTON, EDWARD: born, 1772; probably of the family of Newton, of Bulwell Hall, Nottingham; Ensign, July 6th, 1806; Lieutenant, 1807; volunteered to 66th Foot, August 26th, 1807.

NEVILLE, RICHARD ALDWORTH, of Billingbear: born, 1750; only son of Richard Neville Aldworth, of Stanlake, County Oxon; Lieutenant, 1779; became second Baron

Braybrooke; assumed the surname and arms of Griffin in addition to Aldworth Neville in 1798; married, 178-, Catherine, daughter of Right Honourable George Greenville, M.P. for Reading; died, 1821.

NEVILLE, HONOURABLE RICHARD, of Billingbear; born, 1783: son of Richard Aldworth Neville, second Lord Braybrooke; Captain, 1803; resigned, 1804; M.P. for Berkshire, 1812; married, 1819, Jane, daughter of Charles, second Marquis of Cornwallis. He edited several literary works; died, 1858.

NORREYS, LORD. (See *Abingdon*).

NORRIS, JOHN: born, 1793; perhaps of Hughendon, County Bucks; Ensign, November 17th, 1812; Lieutenant, April 29th, 1814. "Died 1816, at Colchester, Lieutenant John Norris, of the Engineer Company in Fort St. George, East Indies, late of Castle Street, Reading."

NORRIS, H. C., born, 1822; Ensign, April 24th, 1860; Lieutenant, April 16th, 1861.

NOYES, THOMAS BUCKERIDGE, of Southcote: son of G. Noyes, of Andover and Southcote, by Ann, daughter of Charles May, of Basingstoke, whose wife was Anne Noyes, of Southcote; Ensign, 1758; married, 1762, Sarah, daughter of Robert Hucks, of Aldenham, Herts, and Great Russell Street, Bloomsbury, she died at Southcote, 1789. Daniel May, of Sulhamstead, entailed his property on the children of his sisters, Jane, wife of William Thoyts, and Anne, wife of George Noyes; thus the property of Sulhamstead came to the Thoyts family, but if John Thoyts had died it would have gone to his cousin, Thomas B. Noyes. Thomas Buckeridge Noyes was buried at St. Mary's, Reading, November 18th, 1797.

OLDFIELD, CHRISTOPHER CAMPBELL: Half-pay, late 85th Regiment; born, 1838; son of Henry S. Oldfield, Bengal Civil Service; Captain, October 18th, 1873; married, 1872, Edith, daughter of Richard S. Guinness, of Deepwell, County Derby.

OSBORN, WILLIAM : Lieutenant, 1800.

OSGOOD, LAWRENCE HEAD, of Barkham; Lieutenant, 1758; Sheriff of Berkshire, 1747. "We hear, from Newbury, that Lawrence Head Osgood, of Winterbourne, has given twenty guineas to the poor of several parishes in that neighbourhood" (*Reading Mercury*, 1768). He died at Salford, near Oxford, September, 1768, aged 46, and universally regretted.

OTWAY, CHRISTOPHER C.: born, 1838 ; Captain, October 18th, 1873.

OTWAY JOCELYN TUFTON FARRANT, late 5th Dragoon Guards and 49th Regiment: born, 1852; fourth son of Captain William Majoribanks Hughes, who assumed the name of Otway in 1873, by Georgina Frances, heiress of Sir William Loftus Otway; Captain, May 3rd, 1881 ; Hon. Major, June 17th, 1890; Major, May 1st, 1894; resigned as Lieut.-Colonel, March 13th, 1895; married, September 25th, 1884, Eva May, daughter of John Lane Clairmonte, Esq.

PAGE, FREDERICK, of Goldwell House, Speen : Lieut.-Colonel, commanding the 1st Battalion of Local Militia, consisting of ten companies, 1809; died, April 8th, 1834, age 64.

PARKER, JOHN : born, 1791 ; Ensign, April 30th, 1814; Lieutenant, October 6th, 1814; resigned, October, 1852.

PARKER, ROBERT : probably son of Robert Parker, who died 1778, descended from Rev. Thomas Parker, Vicar of Newbury, a Puritan, who emigrated to Massachusetts in 1634 and died 1677. Ensign, 1779; Lieutenant, 1781 ; resigned February 12th, 1787. "Married, May, 1782, at St. James's, London, Robert Parker, to Miss Shelley, of Turville Park, Bucks."

PEACOCK, ——: Major, 1660. Among the estates sequestrated by Parliament and compounded for, occurs the name of John Peacock, of Cumnor.

PEARSON, C. L. M.: born, 1859; son of General Pearson, Assistant Commissioner of Metropolitan Police; Sub-Lieutenant, February 9th, 1876; Lieutenant, February 9th, 1876; went to the Rifle Brigade, December 3rd, 1878.

PECHELL, EDWARD RODNEY CECIL: born, 1840 (*Burke's Peerage* says 1837); son of Captain Samuel George Pechell, R.N., of Bereley, Hants, by Caroline, second daughter of William Thoyts, of Sulhamstead; Lieutenant, April 9th, 1870; Captain, March 27th, 1872; resigned, July 23rd, 1872; was also in 100th Canadian Rifles and the Military Train; married Alice Alleyne, daughter of Rev. John Rothwell; died 1880, leaving two daughters.

PECHELL, WILLIAM MORTIMER CHARLES: born, 1850; nephew to the above and son of William Mortimer Pechell, 85th Regiment, by Georgina, daughter of John Harrop; Lieutenant, February 17th, 1871; Captain, February 5th, 1873; resigned, August 1st, 1883; afterwards went to 3rd Northumberland Fusiliers; married, 1888.

PERCY, LORD, GEORGE ARTHUR MALCOLM: born, 1851; Adjutant, Grenadier Guards, May 9th to April, 1881; then in the Berkshire Militia; transferred as Major to 3rd Northumberland Fusiliers, June 24th, 1886, and now commanding officer; married, 1880, Lady Victoria Frederica Caroline Edgcumbe, daughter of William Henry, fourth Earl of Mount Edgcumbe. He was M.P. for Westminster.

PHILLIPS, JOHN, of Culham: born, at Hagbourne, 1784; son of John Phillips, of East Hagbourne, who married, 1783, Miss Selwood, of Abingdon; Ensign, 1805; Lieutenant 1806; Captain, June 30th, 1806; resigned, 1808; D.L. for Berkshire, 1807; D.L. for Oxfordshire, 1816; High Sheriff for Oxfordshire, 1816; married, in 1809, Frances Anne, daughter of William Cunliffe Shaw, of Singleton Lodge, County Lancaster; she died in 1824; he also died in the same year, and left his property to his nephew, John Phillips. An adventure happened to the Phillips, April,

1799, while travelling in their own carriage from Henley to Maidenhead. They were stopped by two highwaymen at Maidenhead Thicket, good-looking young men, mounted on blood horses. They took away Mr. Phillips' watch and eight guineas in money, but returned the lady's watch most politely. They were afterwards tracked along the road for some way, but took to the woods at Bisham and so escaped capture. Another John Phillips, of Culham, was Carpenter to His Majesty's Board of Works. He died December, 1775, his wife having died the previous October in Reading.

PHILLIPS, WILLIAM : born, at Culham, 1789 (the Militia Register says 1792); son of John Phillips by his second wife, Miss Morland; Sub-Lieutenant, June 26th, 1815; Lieutenant, October 13th, 1825.

PHILLIPS, GERALD EDWIN, of Culham : born, 1870; second son of John Shaw Phillips, of Culham, by Maria Elizabeth, only daughter of Henley G. Greaves, M.F.H., of Newhouse, Abingdon; Second Lieutenant, May 4th, 1889; resigned, February 7th, 1890.

PINKNEY, GEORGE : Ensign, February 25th, 1800.

POCOCKE, JOHN, of Blewbury (or North Fawley in 1786): Ensign, April 1st, 1795 ; Lieutenant, August 4th, 1795.

POCOCKE, JOHN BLAGRAVE, of East Hagbourne: born, 1766 ; Lieutenant, 1795; Captain, May 18th, 1798; Major, November 19th, 1810; resigned, 1825 ; married Charlotte, daughter of John Blagrave.

POLE, HENRY, of Waltham Abbey: born, 1819; Lieutenant, December 30th, 1845; Captain, July 2nd, 1846; resigned, 1853 ; married, 1849, Eliza Anne, daughter of the Rev. Watson William Dickens, Rector of Adisham, Kent.

POPHAM, FRANCIS WILLIAM LEYBORNE, of Littlecote : born, 1862; eldest son of Francis Leyborne Popham ; Second Lieutenant, January 12th, 1881 ; resigned, March 28th, 1882 (never joined) ; succeeded his uncle, 1881.

X

PORTER, FREDERICK, of Whiteknights: Lieutenant, March 20th, 1884 (never joined); transferred to the 3rd Battalion Bedfordshire Regiment, May 5th, 1884.

POWELL, ARTHUR ANNESLEY: Lieutenant, 1792; resigned, 1795.

POWNEY, PORTLOCK PENYSTON, of St. Ives' Place: son of Penyston Powney, M.P., of Windsor, who died, 1758; Captain, 1779; Major, August 31st, 1786; Lieut.-Colonel, April 28th, 1787. He became Colonel eventually, but I have not the dates of his promotion. He married, in 1772, Miss Franklin, daughter of Major Fred Frankland [or Franklin], of the Blues, niece to Sir Thomas Franklin, Bart., a very agreeable young lady with a large fortune; she died the following year, being only 22 years of age, and, in 1776, he married Miss Floyer, of Southcote, but apparently left no son. He was M.P. for New Windsor, and died at the house of his mother-in-law, Mrs. Floyer, in Reading, 1794, in his 57th year. He was the last of the family, and was buried with his father at Windsor.

PRATT, LORD GEORGE MURRAY, of Sunningdale: born, 1843; second son of George Charles, second Marquis of Camden, by Harriett, daughter of the Right Rev. George Murray, D.D., Bishop of Rochester; Captain, November 24th, 1883; Major, May 5th, 1895; formerly Captain, Grenadier Guards, and West Kent Yeomanry. He married Charlotte Harman, eldest daughter of first Lord Cheylesmore.

PRAED, WILLIAM MACKWORTH, of Warfield: third son of William Mackworth (who took the name of Praed, and died in 1752); Captain, 1758. He had a quarrel with Colonel John Dodd in 1762, when he asserted the latter had prevented his promotion; by Court-Martial the suit was decided against him, and, I suppose, he then left the regiment. William Mackworth Praed, of Bitton, Devon, married Susannah, daughter and co-heiress of Thomas

Stokes, of Rill. He was M.P. for Cornwall; and, in 1765, he sold the Manor of Foxley, in Binfield, to Henry Vansittart.

PRESTAGE, THOMAS: born, 1777; Ensign, September 13th, 1811; Lieutenant, January 23rd, 1812; resigned, August 12th, 1812.

PRESTON, JOHN: Sergeant; married, while at Winchester Barracks, June 29th, 1799, Elizabeth Mountain, of St. Thomas; he was 24 and she 22.

PRESTON, JOHN, late —— Militia: born, 1849; Captain, March 28th, 1883; resigned, February 24th, 1888; served local forces—medal—South Africa, also Gold Coast. Resident Magistrate in Ireland.

PRICE, JOHN CHARLES, of The Ham, Farnborough: born, 1747; second son of John Price, by Anne, daughter of Henry Robins, of Wootton Basset; Captain, 1781; matriculated at Brazenose College, Oxon, October 26th, 1764. He was a Justice of the Peace, and Deputy-Lieutenant for Berkshire; died, unmarried, August 11th, 1786, aged 36. Mr. Leonard Price says he died May 11th. (See also page 119.)

PURCELL, JOHN: born, 1779; Ensign, 1805; Lieutenant, March 3rd, 1806; appointed Ensign, 6th Foot; married, Margaret, daughter of Christopher Wyvill, son of Sir Marmaduke Wyvill, by his wife Henrietta Maria, daughter and co-heiress of Colonel Thomas Blagrave, Governor of Wallingford Castle.

PURNELL, EDWARD KELLY: son of E. K. Purnell, Master of Wellington College; Second Lieutenant, July, 1896.

PURVIS, EDWARD, of Darsham, County Suffolk: born, 1786 (Militia Register, 1789); second son of Charles Purvis, of Darsham Hall, Suffolk, by Elizabeth, daughter of Edward Holdon, of Cruttenden; Adjutant, January 16th, 1813; resigned, February 19th, 1846; married, 1817, Lettice

X 2

Elizabeth, daughter and heiress of Rev John Mileso, of
Twywell, Northamptonshire; they lived at Watlington
House, Reading, now the Kendrick Girls' School.  He was
in the 4th Foot, fought in the Peninsular War, and was
wounded at the Battle of Corunna.  There is an account
of Captain Purvis in Mr. Darter's interesting *Reminiscences.*

PYE, WALTER, of the Temple, London: Lieutenant,
1779; Captain, 1781 ; Major, October 22nd, 1793.  He was
the son of Henry Pye, of Faringdon, M.P. for the County of
Berks, by Mary, daughter of Rev. David James, Rector of
Wroughton, Buckinghamshire; she was the only sister
of Anthony James (James, of Denford, near Hungerford),
who assumed the name of Keck, in 1737.

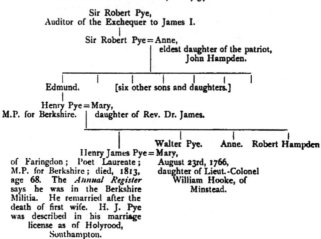

Walter Pye died unmarried, at Carmarthen Street, Bedford Square, London, on January 9th, 1824, age 72.  He
was Senior on the list of Commissioners of Bankruptcy.
Henry Pye was elected five times Member for Berkshire
without opposition; he died, 1766, just as his eldest son

Henry James Pye, came of age. This last gentleman injured his fortune so materially in a contested election for Berkshire, that, in 1784, he was obliged to sell the paternal estate; he then was appointed Poet Laureate, and he was a Police Magistrate of Westminster. The family was said to have come over with William the Conqueror, and settled in Herefordshire, and from thence to Faringdon.

RADNOR, SECOND EARL OF, JACOB: born, 1750; Lord Lieutenant of Berkshire from 1791; Colonel of the Berkshire Militia, December 24th, 1791; resigned, May 30th, 1801; married, Hon. Anne Duncombe, daughter and co-heiress of Anthony, Lord Faversham, so says Burke, but the *Reading Mercury* of 1776 gives the following entry: "Married, at St. George's, Hanover Square, the Earl of Radnor to Miss Duncombe, daughter of Thomas Duncombe, Esq., Member for Downton, Wilts"; died, 1828. His portrait may be seen in the Town Hall, Wallingford.

RAMSEY, ———; Lieutenant, June, 1798.

RAVENSHAW, THOMAS WILLIAM, of Bracknell: Ensign, February 20th, 1794; Lieutenant, August 1st, 1794; Captain, October 18th, 1796; Lieut.-Colonel, August 21st, 1800; Colonel, December 9th, 1812; died, August 14th, 1842. His name disappears from the *Army List*, 1842. "1762, married, Thomas Ravenshaw, Esq., merchant, to Miss Ann Wilmott, of Old Jewry"; this was probably his father's marriage. In 1803, Colonel Thomas W. Ravenshaw took out a shooting license as "of Easthampstead."

REED, H. WILSON, M.D.: Surgeon, February 24th, 1855; resigned, 1878 (?), since which time the Regiment has not had its own Regimental Surgeon.

REID, GEORGE ALEXANDER CARADOC: Lieutenant, March 15th, 1873; resigned, November 20th, 1873; after-wards in the Queen's Own Cameron Highlanders (formerly 79th Regiment), as Lieutenant, November 20th, 1875,

Captain, September 19th, 1881. (Lieut.-Colonel George Alexander Reid was M.P. for New Windsor, 1845, and died, 1852. In 1773, Rev. Mr. Reid bought White Waltham Place, but sold it again in 1776.)

REEVES, EDWIN, of Arborfield: born, 1756; Lieutenant, January, 1794; Captain, May 27th, 1797; resigned, June 16th, 1806; married, 1805, "Edwin Reeves to Miss Warner, eldest daughter of John Warner, of Beaulieu, Hants."

REEVES, JOHN, of Arborfield: Captain, 1758; married Dorothy, daughter of Charles Gore, of Hackston, County Lincoln, sister and heiress of Pelsant Reeves, Captain 1st Royals, who was killed at Toulon, 1793; died at Andover, 1813, in his 80th year. He was verderer of Windsor Forest. In 1774, a servant of his was stopped by two highwaymen between Guildford and Frimley, who robbed him of 2½ guineas in gold and some silver.

RICKMAN, STUART HAMILTON: born, 1873; son of Colonel Albert Rickman, late of Rifle Brigade, living at Inkpen, near Newbury; Lieutenant, January 24th, 1891; went to the 3rd Battalion Rifle Brigade, November 29th, now in 1893; the Punjaub.

RHODES, JOHN EDWARD, of Hennerton: eldest son of John William Rhodes, of Hennerton (a Lieutenant in 60th Rifles), by Marie Ada, eldest daughter of Edward Mackenzie, of Fawley Court; Second Lieutenant, June 7th, 1887; Lieutenant, November 17th, 1888; went to King's Royal Rifle Corps (60th Rifles), October 29th, 1889; married, February 18th, 1897, at St. John's Church, Ryde, Beatrice Zoe, youngest daughter of Sir Richard Sutton, fourth baronet, of Benham, County Berks.

RHODES, HUBERT VICTOR, of Hennerton: born, 1874; brother of the above; Second Lieutenant, February 8th, 1892; Lieutenant, February 20th, 1895.

ROBBINS, JOHN: born, 1783; Ensign, June 6th, 1809; Lieutenant, August 22nd, 1809; resigned, November 17th,

1810; Lieutenant Half-pay 44th Regiment, 1810. Probably of the Hampshire family; he had a son and heir born 1815. In 1802, John Robbins, 4th Foot, age 21, married at Portsea, Anne Collins, of Petersfield.

ROE, WILLIAM: born, 1794; Lieutenant, February 22nd, 1815. "Married, 1775, William Roe, Esq., to Miss Thomas, daughter of Sir William Thomas, of Yapton, Sussex." There was a Rev. John Roe, Vicar of Newbury, 1797; he came from Macclesfield.

ROUND, HENRY: born, 1786; Ensign, 1807; volunteered the same year to the 63rd Foot. Some land in Swallowfield belonged to the Rounds, one of whom was a lawyer in Windsor, and died at Foster House, Egham, 1804. Another of the family, Henry Round, was buried at Henley in 1821, age 71; he lived at Abney House, Woburn. Stephen Round lived at Beech Hill.

ROUSE, RICHARD: born, 1774; Ensign, December 26th, 1805; resigned, 1807,

ROYDS, ALBERT HENRY: born, at Brownhill, Rochdale, April 7th, 1876; eldest son of Edmund Albert Nuttall Royds, of Falinge, Lancashire, by Augusta Eliza, daughter of A. H. Lemonius, of Stonehouse, County Lancaster; educated at Eton; Second Lieutenant, August 3rd, 1894; Lieutenant, May, 1895. Mrs. Royds rents Standen Manor, near Hungerford.

RUDLAND, JONES: born, 1785; Ensign, September 1st, 1809; Lieutenant, November 28th, 1809; volunteered to 10th Foot, May 1st, 1811.

SAVERNAKE, VISCOUNT, GEORGE WILLIAM THOMAS: 2nd Lieutenant, January 29th, 1881; resigned, April 11th, 1886; married an actress; died, 189-.

SAUNDERS, LIEUT.-COLONEL, 1667: probably one of the Saunders of Chaddleworth. One branch of the family lived at Sulhamstead and many are buried in Meales Church.

SAUNDERSON, ANTHONY: born, 1762; Ensign, 1806; Lieutenant, 1807; displaced, 1808. He may have been a cousin of the Vansittarts. There was an eminent architect named Saunderson, who died in 1812.

SAWYER, JOHN: born, 1762; eldest son of Anthony Sawyer, of Heywood, by his second wife, Phœbe, daughter and co-heir of Richard Harcourt, of Wigsell, County Sussex: married, 1785, Sarah, daughter of Anthony Dickins, of Cherrington, County Warwick; Lieutenant, February 12th, 1787; resigned, May, 1790; died, 1845.

SAWYER, CHARLES, of Heywood, White Waltham: born, 1813; son of Charles Sawyer, of Heywood, by Henrietta, eldest daughter of Sir G. Bowyer, Bart.; Ensign, March 16th, 1831; Captain, December 23rd, 1831; Major; went to the Line, 16th Light Dragoons; became Lieut.-Colonel 6th Dragoon Guards; married Anna Maria, daughter of T. J. Timins, of Hilfield, Herts.

SCHRADER [or SCHRODER], FREDERICK HENRY: Ensign, December 19th, 1798; Lieutenant, February 5th, 1799. "In 1749, John Adolph Schroder married Miss Anne Mighello."—*Universal Magazine.* "1781, at St. George's, Hanover Square, Henry Otto Schrader to Virginia Louisa Forster, of Paddington."

SCOTT, HENRY FARQUHAR: born, 1871; second son of Sir Edward H. Scott, of Lychet Minster, County Dorset, by Emilie, daughter of Lieut.-Colonel Packe, of Twyford Hall, County Norfolk; 2nd Lieutenant, 1896; Lieutenant, April 7th, 1897. In the Charter Company Police (had a narrow escape, December, 1895, being in the advance guard of Jameson's celebrated ride across the Transvaal).

SELLWOOD, RICHARD, of Peasmore: Lieut.-Colonel, 1771; died, at Brightwalton, October, 1776. His only son died a few weeks before him; his wife died, December, 1771, at Peasmore, universally respected for her humane and sweet disposition. There was another family of

Sellwood at West Illsley, one of whom married in 1775 the daughter of John Rowland, of Aldworth, an accomplished young lady with a fortune of £10,000. There were also Selwoods of Abingdon (see page 260), Welford, and Aldworth. All these were probably descended from the same.

SEXTON [or SAXTON], CLEMENT, of Caldecott: Captain, 1762; Major, ———; Lieut.-Colonel, 1781; resigned, August 28th, 1787. A trial took place at Abingdon, anent bribery in the election of 1768, between Captain Sexton and Mr. Sherwood, a carrier. It was decided in favour of the plaintiff, Mr. Sexton. He was High Sheriff for Berks, 1778, and was brother to Sir Charles Sexton, of Circourt, Abingdon.

SEYMOUR, ROGER, of Inholmes, Lambourn Woodlands: Lieutenant in the Trained Bands; died, 1631.

SEYMOUR, EDMUND, of Inholmes, Lambourn Woodlands: Lieutenant, 1758; Captain, 1779; died, 1798, and buried at Lambourn. "Edmund Seymour, of Lambourn, age, 30; 1781; Catherine Draper, of St. Swithin's, Winchester, age, 21, spinster."

SHAW, LIEUT.-COLONEL: Paymaster; resigned, December 27th, 1858.

SHACKEL, WILLIAM, of Basildon: born, 1787; Ensign, July 21st, 1812; resigned, March 14th, 1813. There is a monument to him in Basildon Church.

SHACKEL, WILLIAM RICHARD, of Mile House, Sulhamstead; son of William Shackel, who rented Mile House and Lower Basildon Farm; Ensign, November 30th, 1852; Lieutenant, May 27th, 1854; Captain, April 16th, 1861; resigned, May, 1878.

SHAW, PONSONBY, LIEUT.-COLONEL: Paymaster, 1855.

SHERREN, WILLIAM: born, 1781; Ensign, February 4th, 1799; Lieutenant, May 18th, 1799; Full pay, 43rd Foot,

1803. The Militia Register gives him as ensign, 1803, but I think it is wrong unless there were two William Sherrens.

SHERSON, ALEXANDER NOWELL: born, 1816; youngest son of Robert Sherson, by Catherine, daughter of Captain John Taylour, 72nd Highlanders; served in the Kaffir War, then Half-pay Adjutant Royal Berks Militia, February 19th, 1846; Captain, October 17th, 1852; resigned, February 7th, 1855; married, 1854, Lady Anne Maria, daughter of John, fourth Marquis of Townsend; died, 1882.

SHEPPARD, EDWARD, of Speenhamland: Adjutant, 1777; Captain-Lieut., 1779; Captain, 1781; Captain of a Company, November 17th, 1786; resigned, February 1st, 1796. He commanded the Newbury Volunteers, consisting of 5 officers and 104 men, July 26th, 1799, when reviewed on Bulmarshe Heath.  Married, August, 1777, at Newbury, Miss Gale, of Lackhamstead, "a most amiable young lady possessed with every accomplishment to insure happiness in the marriage state." "Died, September, 1800, at his house at Speenhamland, Edward Sheppard, Deputy Lieutenant for Berks, and for many years Adjutant of the Berkshire Militia. His rich vein of anecdote and inexhaustible flow of spirits, which were wont to set the table in a roar (notwithstanding a lingering and painful disorder which at last terminated in his dissolution in the 60th year of his age,) will occasion his loss to be severely felt by his numerous friends and acquaintances."

SIDNEY, PHILIP: born, 1787; fourth son of John Sidney, of Court Lodge, Yalding (who claimed the title of Earl of Leicester), by Elizabeth, daughter of Rev. J. Apsley, of Ripple, Kent; Ensign, October 3rd, 1809; resigned to 43rd Foot, June 5th, 1810; married, Sophia Everett.  His brother, Sir W. R. Sidney, lived at the Bourne, Maidenhead.

SIMPSON, FRANCIS: Ensign, May 24th, 1790; Lieutenant, October 19th, 1792; resigned, 1793.

SLADE, WILLIAM A. (possibly of Lockinge): born, 1776; Ensign, 1804; Lieutenant, November 27th, 1805; resigned, 1808. William Slade, of Wallingford, married at Basildon, in 1704, to Mary Tull, was probably an ancestor.

SLADDEN, WILLIAM, of Chatham, Kent: Captain, 1781; married, 1775, at St. Mary's, Reading, to Elizabeth Russell.

SLOCOCK, CHARLES SAMUEL, of Donnington: born, 1821; eldest son of Charles Slocock, banker, of Donnington, by Sophia, only daughter of Silas Palmer, of Isleworth; Captain, October 11th, 1852; resigned, 1861; married, 1860, Mary, daughter of Edward Goddard, of Stonehouse, Berks. He sold Donnington, 1896, to Colonel Downes.

SLOCOCK, EDMUND: born, 1787; son of Samuel Slocock, of Newbury, by Mary, daughter of John Merriman, of Speen; Ensign, September 9th, 1815; Paymaster, March 15th, 1816; resigned, May 1st, 1832; married Elizabeth, daughter of Henry Greenway, of Cassington, County Oxon.

SMITH, BENJAMIN: Ensign, June 5th, 1799.

SMITH, JAMES WILLIAM, of Kidwells, Maidenhead: born, May 3rd, 1833, at Maidenhead; son of James Smith, of Kidwells, by Elizabeth, only daughter of William James Jones, of the Forge House, Egham, Surrey, stockbroker; gazetted Ensign, May 4th, 1855; joined at once; Lieutenant, November 13th, 1857; left, March 12th, 1858; then joined the Maidenhead (5th Company) Berks Rifle Volunteers, of which Colonel Robert Vansittart, of Chuffs, Holyport, was Captain; promoted Sergeant-Major of that Company, March 6th, 1860; married, September 22nd, 1863, Miss A. E. Moore, second daughter of Francis Moore, M.D., of Much Hadham, County Herts. Present address, 15, Woodhurst Road, Acton, S.W.

SNOOK, THOMAS: Ensign, February 15th, 1799.

SOUTHBY, SAMUEL, of Chieveley: Lieutenant, 1762; married, daughter of John Blandy, of Chaddleworth. The

Southbys held the Manor of Appleton from 1577, when John Southby purchased it from Sir John Fettiplace.

SPEED, D W. H.: born, 1760; Ensign, 1808; Lieutenant, 1808; displaced, 1809.

SPEEN, JAMES: Ensign, February 11th, 1791.

SQUIRE, WILLIAM: born, 1778; Ensign, 1808; Lieutenant, 1809; volunteered to 4th Foot, April, 1809, and was a Lieutenant in that regiment, 1811; fought at Waterloo. In 1768, Joseph Squires, of Eton, married Elizabeth, daughter of Joseph Bagnall, of Eton.

STARCK, CHARLES GEORGE: Ensign, 1779. Sigismund Baron de Starck married Martha, third daughter of Nathaniel Ogle, of Kirkley, who, in 1708, had married Elizabeth, daughter of Jonathan Newton, of Newcastle-on-Tyne. Lieut.-Colonel Starck was in command of the Royal York Rangers at Guadaloupe, 1815, mentioned in dispatches. Starck, Von, John Augustus: a German divine and Theological writer, born at Scherverin, 1741, preacher in 1781 to the Court at Darmstadt, so esteemed by the Landgrave of Hesse he made him a Baron; he died, 1816. In 1776, Charles Sigismund Baron de Starck, died at Oxford, his widow, Martha, Baroness de Starck, died in Baker Street, London, 1805, age 86.

STARES, WILLIAM, of Gosport: born, 1830; Ensign, January 16th, 1855; Lieutenant, January 8th, 1856; resigned, February, 1859. Probably came into the Militia through Major John Leveson Gower, who lived at Gosport. In 1791, William Stares, of Bishops Waltham, married Susanne Gater, of North Stoneham.

STEAD, FRANCIS SACHEVERILL, of Donnington: Major; Commanded the Troop of Horse, really the old Militia Horse Troop, called the Reading Provisional Cavalry, which went to Ireland in 1798. He died, 1826, age 67.

STEPHENS, HENRY: born, 1793; Ensign, June 13th, 1811; Lieutenant, January 22nd, 1812; gazetted to 14th Foot, December 25th, 1814.

STEPHENS, HENRY GILBERT: born, 1752; Ensign, August 19th, 1793; Lieutenant, December, 1793; Paymaster, 1794; resigned, May 24th, 1815; left about 1820. These Stephens were evidently of Catmore. John Stephens, of Peasmore, married the only child of Colonel Richard Selwood, of the Berkshire Militia. Both are buried at Peasmore.

STEPHENSON, ROWLAND, of Farley Hill. He took a troop of eighty men to oppose the Spanish Armada in 1539 (so says Burke); but, I think the date ought to be 1589!

STEPHENSON, WILLIAM: Lieutenant, H Troop, August 23rd, 1650.

STEPHENSON, JOHN, of Farley Hill (now represented by Standish, of Standish, County Lancaster): Lieutenant, 1779.

STEWART, FRANCIS T.: Lieutenant, March 6th, 1885; went to Highland Infantry, November 9th, 1886.

STONOR, ——: Cornet in the Wallingford Troop of the King's Army; killed at the siege of Basing House, 1644.

STRACEY, THOMAS, of London: Ensign, 1794; Lieutenant, January, 1795; resigned, 1796. Thomas Stracey, a London Merchant, died at his house at Wallingford, 1773, perhaps father of the above.

STRATTON, WILLIAM: born, 1787; Ensign, January 19th, 1813; left about 1820. The daughter of William Stratton married John Hector Cherry, of Denford.

STUART, JOHN: born, 1783; Ensign, 1807; Lieutenant, 1807; resigned, 1808; Half-pay, 1808, 71st Foot.

STURGES, GEORGE TRAVIS: born, 1778; Ensign, 1803; Lieutenant, 1803; Captain, February 13th, 1805; died,

October 2nd, 1808. Perhaps son or grandson of Rev. C. Sturges, Vicar of St. Mary's, Reading. Died, October 2nd, 1808.

STURGES, JOHN: born, 1779; brother to the above; Ensign, 1803; Lieutenant, 1803; Captain, 1805; resigned, 1828.

STURGES, S. H.: born, 1842; Lieutenant, January 11th, 1862.

STYLES, CLEMENT: Ensign, 1762. Sir Clement Styles married in 1794.

SUDLEY, VISCOUNT ARTHUR JOCELYN CHARLES: born, 1868; son of the Earl of Arran by Hon. Edith, second daughter of Viscount Jocelyn and grand-daughter of Robert, third Earl of Roden; Second Lieutenant, May 14th, 1887; Lieutenant, November 17th, 1888; went to the Royal Horse Guards, November 19th, 1889.

SYKES, SIR FRANCIS, BART., of Basildon: born, 1767; son of Francis Sykes, of Acworth Park, who died January 12th, 1804, Governor of Cossimbogar, Bengal, where he made a fortune and, on his return from India, 1770, bought the Manor of Basildon, including the Grotto, being the settled estate of the late Viscount Fane, for £47,000, and the unsettled estate for £12,000, from the Countess of Sandwich and Salis. I fancy the family belonged previously to Reading, as the name occurs frequently in the accounts of the Reading Corporation. His father must have become a widower and remarried, as in 1774 (by special license) the wedding took place, at Little Sion House, between Francis Sykes, of Basildon, and Miss Elizabeth Monckton, daughter of William Viscount Galway. This lady and her husband were shortly afterwards presented at Court in St. James's Palace. Her jewels on this occasion were very beautiful. She wore a suit of pearls, the finest in England (the Queen's only excepted) and her diamonds were remarkable for their beauty and magnificence. Francis Sykes, the son, was first and foremost in all county business in Berkshire and was M.P. for Wallingford, 1784-1802. He married, 1798, Mary Anne, eldest daughter of Hon. Major Henniker, grand-daughter of John, first Lord Henniker. They had several children. While on a visit to Elberfield, in Germany, with their children, the wife caught the scarlet fever and died February 27th, 1804. Her husband, who had nursed her with great devotion, caught the fever also and died March 7th, 1804, only having outlived his father a few weeks. Their bodies were conveyed to England. They were landed at London, and a State procession was formed. In the first hearse was the body of Sir Francis Sykes, then followed a hearse with the body of Lady Sykes and her infant child. Three mourning coaches, drawn by six horses, came next, then the family coach; Lady Sykes' brother in a coach drawn by four horses, and lastly, many of the tenants on horseback. On arriving at Basildon, the bodies lay in state and were afterwards interred in the family vault. The four little orphan children did not form part of this sad and solemn *cortège*, but came to England after the funeral was over.

TEBBOTT, ROBERT, of Windsor: born, 1832; Ensign, May 17th, 1855; Lieutenant, May 19th, 1859. Robert was a burgess of Windsor in 1813, and another member of the family led the band at the Reading Subscription Concerts in 1785.

TEMPLE, CHARLES PILCHER, 49th Regiment: Adjutant, September 20th, 1880; rejoined 1st Battalion, January 30th, 1885, as Major, commanding his Battalion; afterwards commanded Regimental District, Worcester; now Assistant Adjutant-General, Cork; married Rosa, daughter of Admiral Bonham; she died, 188–.

THEOBALD, FREDERICK CAMBRIDGE, of Sutton Courtney: born, February 25th, 1864; second son of Theobald

Theobald, of Sutton Courtney Abbey, by Elizabeth, daughter of John Justice, of Sutton Courtney; Lieutenant, March 10th, 1883; went to 2nd Battalion, November 12th, 1884; was in the Egyptian Transport Service, and finally in the West Riding Regiment; resigned, March, 1895; married, 1893, Emmeline, daughter of Robert Goodson, of Barkston Towers, Tadcaster. Captain Theobald bought Sutton Courtney House, 1895.

THOMAS, STEPHEN: appointed Muster Master of Reading, 1615.

THOMSON, EDWARD TEW: Lieutenant, October 6th, 1852; Captain, January 20th, 1848; resigned, February 19th, 1863; married Miss Lightfoot.

THORNTON, WALTER, of Maidenhatch: son of C. J. Thornton, formerly of St. Petersburg, now of Meran, Tyrol; Second Lieutenant, June 20th, 1887; Captain, May 4th, 1889; Instructor of Musketry, 1891; passed as interpreter of the Russian language; married twice, his second wife is Gertrude May Sturges, daughter of Rev. Simon Sturges, of Wargrave. He rented Calcot Grange, and in 1894 purchased land at Maidenhatch, in Pangbourne, and built a house there.

THOROWGOOD, SIR JOHN, of Billingbear, 1646.

THOYTS, MORTIMER GEORGE, of Sulhamstead: born, November 6th, 1804; only son of William Thoyts, of Sul-hamstead, by Jane, daughter and co-heiress of Abraham Newman; Sheriff for Berks; Captain, June 15th, 1832; resigned, March 13th, 1833. Married thrice: 1828, at Padworth, Emma, daughter of Thomas Bacon, of Aberavon, South Wales, who rented Benham and Padworth, and afterwards lived at Redlands, Reading; 1848, Catherine, daughter of Robert Sherson, of Fetcham, and widow of Captain Smith, of Tilehurst; 1872, Catherine, daughter of — James, Esq., and widow of Rev. Robert Sherson, of Yaverland, Isle of Wight. M. G. Thoyts died, January,

1875, and was buried in Meales Churchyard, Sulhamstead. He was presented by the electors of Berkshire, for the work he had done politically, although he refused to represent the county in Parliament, with a fine portrait of himself, painted by J. Horsley, R.A.

THOYTS, WILLIAM RICHARD MORTIMER, of Sulhamstead: born, December 29th, 1828; eldest son of Mortimer George Thoyts by his first wife, Emma Bacon; Captain, October 9th, 1852; Second Major, April 15th, 1861; resigned, after two years' serious illness, January 22nd, 1873; married, 1856, at St. George's, Hanover Square, Anne Annabella, eldest daughter of Sir Richard Puleston, Bart., of Emral. He was High Sheriff for Berks, 1883.

THROCKMORTON, JOHN PHILIP HOWARD, of Buckland: born, April 14th, 1840; fourth son of Sir Robert George Throckmorton, by Elizabeth, only daughter of Sir John Acton, Bart., of Admaston, County Salop; Lieutenant, May 10th, 1869; resigned, October 5th, 1872.

TIMBRELL, WILLIAM HALL, of Streatley: Lieutenant, 1781; Captain, February 9th, 1787; resigned, 1795; married, 1784, at Marylebone Church, London, to Miss Nash, of Sevenoaks; died, at Lewisham, aged 62. He owned property at Sevenoaks through his wife.

TOLL, ASHBURNHAM NEWMAN: third son of Ashburnham Toll, of Preston Deanery, County Northampton, by Mary, daughter of Lieut.-Colonel Geary (his son, the Rev. Ashburnham Philip Toll, Prebendary of York, born 1743, assumed the name of Newman in December, 1775); Ensign, April 24th, 1879; Captain-Lieutenant, May 17th, 1789; Captain, January, 1795; married, 1773, Mary, daughter of Paul Mowbray, of London, relict of Captain Alexander Wood, but died without issue, March 16th, 1802. This puzzles me, as apparently it was the clerical Ashburnham Toll who was in the Militia, as his nephew, Ashburnham Cecil Newman Toll, was not born till 1796. He resided in

Y

Reading in 1786, as his gun license was taken out as of that town.  Ashburnham Toll, of Greywell, Hants, married Anne, daughter of Richard Newman, of Evercreech Park, County Somerset.

TOOGOOD, CHARLES FREDERICK STRANGEWAYS GLYNNE, of Dean Wood: born, 1861; eldest son of Octavius Toogood, late of Indian Civil Service, of Dean Wood, by Clara, daughter of Commander Lawrence Gwynne, of Cambrian, County Devon; Second Lieutenant, March 26th, 1879; Lieutenant, May 10th, 1880; services dispensed with, 1883.

TOWSEY, WILLIAM (the younger), of Wantage: Ensign, 1758.

TREACHER, GEORGE: only son of Rev. Thomas Treacher, of Audley; educated at Eton; Ensign, 1798; Lieutenant, 1798; went to 2nd Life Guards, 1800; married, 1825, at Sonning, to Mary, youngest daughter of Mr. Bullock, of Hampstead Farm, Sonning.  I suppose he married twice; as I find George Treacher, the husband of Harriet Rachel, daughter of Alan Swainston, M.D., of York, by his wife Frances, heiress of Francis Strangeways, of Alne, County York.

TREVOR, HENRY: born, 1783; Ensign, 1807; Lieutenant, displaced, 1808.  It may have been a relation who died, 1784, at her house at Long Wittenham, Mrs. Trevor, relict of Tudor Trevor, and sister of William Jennens.

TRISTRAM, LANCELOT SHUTE BARRINGTON, of Fowley, Hants: born, 1857; eldest son of William Barrington Tristram (who died, 1877), by Eliza Elizabeth, daughter of Rev. Lancelot Miles Halton, of Monck Sherborne; Lieutenant, June 20th, 1877; resigned, February 26th, 1879; went to the Welsh Regiment.

TURNER, CHARLES, of 4, Berkeley Avenue, Reading; Captain, 1st Battalion, January 23rd, 1889; Adjutant, 3rd Battalion, October 31st, 1895; served in the Egyptian

Campaign, 1882 (medal and Khedive's star). He married twice: 1886, Ella, daughter of C. J. Thornton, of St. Petersburg, who died, 1887, buried at Tilehurst; secondly, Janey, daughter of Vice-Admiral Sir Alexander Buller, K.C.B., of Erle Hall, Plympton.

TYLDEN, OSBORNE, of Torry Hill, Milstead: second son of Rev. Richard Osborne Tylden, of Milstead, by Deborah, daughter and heiress of Daniel May; Ensign, 1781; married Anne Withers, of London; they had nine children, the eldest of whom, Osborne, died young. He died in 1827.

UFFINGTON, VISCOUNT, WILLIAM AUGUSTUS FREDERICK: born, 1838; Captain, Grenadiers Guards; appointed Major, March 14th, 1864; died, April 19th, 1865.

URLWIN, JOHN: born, May 4th, 1875; only child of John Urlwin, of the Bungalow, Burghfield; Second Lieutenant, January, 1897.

VAN DE WEYER, WILLIAM VICTOR BATES, of Kingston Lisle and New Lodge, Windsor Forest: born in Portland Place, at his grandfather's house, November 20th, 1839; eldest son of His Excellency, Sylvan Van de Weyer, of New Lodge, Windsor, Envoy Extraordinary and Minister Plenipotentiary to the King of the Belgians, by Elizabeth, daughter of Joshua Bates, of Portland Place; joined, 1862; Major, May 4th, 1871; Lieut.-Colonel, February 22nd, 1881; resigned, with rank of Colonel, April 17th, 1886; married, 1868, Lady Emily Georgina, daughter of William, second Earl of Craven.

VAN DE WEYER, WILLIAM JOHN BATES, of New Lodge: born, 1871; eldest son of the above; Second Lieutenant, January 1st, 1890; Lieutenant, February 7th, 1891; Captain, May 13th, 1895.

VAN DE WEYER, BATES GRIMSTON, of New Lodge: younger brother of the above; Second Lieutenant, June 7th, 1894; gazetted to the Scots Guards, January, 1897.

VANSITTART, ARTHUR, of Shottesbrook ; Lieut.-Colonel, 1759; Colonel, April 15th, 1762; married Hon. Anne Hanger, daughter of Gabriel Lord Coleraine; died, 1804.

VANSITTART, ARTHUR, of Shottesbrook: born, 1762; son of the above; Ensign, May 7th, 1798; Captain, July 6th, 1798; Lieut.-Colonel, February 18th, 1801; resigned, September 25th, 1812; married, 1806, Hon. Caroline, fourth daughter of William, first Lord Auckland. Died, June, 1782, at her house at Littleton, County Middlesex, Mrs. Vansittart, mother of Colonel Arthur Vansittart, of the Berkshire Militia, and a few days later, at Shottesbrook, Mrs. Vansittart, that gentleman's wife. Both father and son represented Berkshire in Parliament.

VANSITTART, HENRY, of Kirkleatham, County York: born, 1776; only son of Henry Vansittart, by Catherine Maria Powney; Ensign, October 5th, 1803; resigned, 1803; married Teresa, second daughter of Charlotte Viscountess Newcomer and relict of Sir Charles Turner.     •

VANSITTART, GEORGE HENRY, of Bisham: born, 1768; son of George Vansittart, of Bisham Abbey, by Sarah, daughter of Sir James Stonhouse, Bart.; joined as Major, July 15th, 1798; Lieut.-Colonel, August 24th, 1799; Colonel, May 30th, 1801; Brigadier-General in the Leeward Islands, 1801; Lieut.-General, 1810; then of the 12th Battalion of Reserve as Major-General; married, 1818, Mary Anne, daughter of Thomas Copson; died, 1824. He was M.P. for Berkshire in 1852.

VANSITTART, HENRY, of Shottesbrook; 'born, 1778; Ensign, 1803; Lieutenant, 1804; resigned, 1805; Rear-Admiral of the Red; died, 1843.

VANSITTART, NEALE HENRY, of Bisham: Second Lieutenant, June 15th, 1878; Lieutenant, June 25th, 1879; went to 72nd Highlanders, October 25th, 1880.

VELLEY, THOMAS: Ensign, 1779; afterwards Lieut.-Colonel of the Oxfordshire Militia; lived in Bath. He or his son is mentioned in Miss Anna Seward's *Letters* as "Major Velley." He died of concussion, caused by jumping from a runaway coach from the "Castle Inn," Reading, while on his way back to Bath, 1806. His wife, who was with him, escaped without injury. He was buried at St. Mary's, Reading, June 13th.

VELLEY, CHARLES : born, 1782; probably son of Thomas Velley ; Captain, November 28th, 1809; resigned, February 18th, 1831.

VERE, LORD, SECOND BARON VERE OF HANWORTH : born, 1699; third son of the first Duke of St. Albans; created Baron previous to 1750; Lord-Lieutenant of Berkshire, 1771; died, 1781.

VINCENT, HENRY WILLIAM, of Lily Hill, Bracknell: son of Henry Dormer Vincent, by Isabel, daughter of Hon. Felton Hervey; Captain, June, 1828; resigned, April 7th, 1852; married Elizabeth, daughter of Colonel George Callander, of Craigforth and Ardkinlas; died, 1865.

VINCENT, CHARLES EDWARD HOWARD, of Donnington : born, 1849; second son of Sir Frederick Vincent, Bart., by his second wife, Maria Copley, daughter of Robert Herries Young, of Anchenskrugh, County Dumfries ; married, 1882, Ethel Gwendoline, second daughter of George Moffatt, of Goodrich Court, County Hereford, niece of Mr. Morrison, of Basildon Park; late Lieutenant, 23rd Regiment; Captain, October 18th, 1873; resigned, November 10th, 1875 ; afterwards commanding Central London Rifle Rangers, and now the Queen's Westminster Volunteers; appointed Chief of the Criminal Investigation Department, Scotland Yard ; M.P. for Sheffield ; Knighted, 1893 ; Barrister-at-Law ; Member of French Bar. Sold Donnington in 1894.

VINCENT, EDGAR : born, 1857; Sub-Lieutenant, May 3rd, 1875; went to the Coldstream Guards, December 30th, 1877; Knighted, 188-; President of the Ottoman

Bank at Constantinople; married Lady Helen Duncombe, daughter of the Earl of Feversham.

VOULES, CHARLES STUART, of Windsor: born, 1832; son of Charles Voules, of Windsor; Ensign, March 24th, 1855; Lieutenant, October 2nd, 1857. He is mentioned in *Leaves of My Life*, by Montagu Williams.

WADLING, W. A.: son of Lieut.-Colonel Wadling, late of the 1st Battalion Northumberland Fusiliers; Second Lieutenant, February, 1896; Lieutenant, April, 1897.

WALLIS, JOHN: Ensign, May, 1779; Lieutenant, 1779; resigned, 1793; died, at his house at Kennington Cross, Surrey, October 6th, 1802, age 50. He was in H.M. Customs.

WALKER, JOHN: born, 1788; Captain, April 29th, 1814; dismissed, September 15th, 1815.

WALKER, CHARLES HOULDON: Ensign, March 11th, 1799.

WALTER, JOHN ABEL, of Farley Hill: (his age is given as 39, but only his appointment as Adjutant is noted in the Militia Register, unless both father and son were in the Regiment, which I expect was the case); Lieutenant, 1758; Major, 1779; Adjutant, 1786; Lieut.-Colonel, August 31st, 1786; resigned, June 25th, 1803; married Newton, only daughter of Alexander Walker, of Swallowfield and Barbadoes, she died in 1772. In 1774, John Walter, of Paternoster Row, married Susannah, daughter of Mr. Lambert, a considerable farmer at Warnash, near Guildford. The Walters of Bearwood are descended from the above, who, I fancy, is the John Walter given in Burke's *Landed Gentry* as of Warwickshire, whose marriage is not given.

WALTER, JOHN ABEL: Lieutenant, November 20th, 1786; Adjutant, 1786; Captain by Brevet, October 17th, 1786. John Walters, of Penthygerent, County Cardigan, was the son of John Walters, of Penthygerent; married Frances Griffiths, of Llwyn-y-brain, County Carmarthen.

Their son was Abel Griffiths Walters, who married, 1780, Bridget, sister of Sir T. Philipps, of Abergavenny. Another John Walter married Katherine, daughter of Peter Noyes, of Trunkwell.

WALTER, JOHN BALSTON, of Bearwood: born, 1845; son of John Walter, of Bearwood, by his first wife, Emily Frances, daughter of Major Henry Court, of Castlemans, Berkshire; Lieutenant, June 25th, 1865. Drowned in the lake at Bearwood while gallantly trying to save his brother, December 24th, 1870.

WANTAGE, BARON ROBERT LOYD LINDSAY, V.C., K.C.B., of Lockinge: Captain and Lieut.-Colonel Scots Fusilier Guards, Crimea; Lord Lieutenant of Berkshire, 1881; late Colonel of the Berkshire Volunteers, and Brigadier-General Home Counties Volunteer Brigade; married Miss Loyd, daughter and heiress of Lord Overstone.

WARNEFORD, FRANCIS: Lieutenant, February 10th, 1787; resigned, October 19th, 1792. No doubt related to Rev. Samuel Wilson Warneford, brother of Colonel Warneford, of Warneford Place, Wilts, who married, 1796, Margaret Loveden, daughter of Edward Loveden Loveden, of Buscot.

WAYLAND, JOHN THOMAS [or WEYLAND], of Hawthorn Hill: born, 1784; Ensign, March 3rd, 1806; appointed Ensign 53rd Foot. Possibly of the Oxfordshire family.

WATSON, —: Ensign, 17 ; Lieutenant, 66th Regiment, 1783.

WEEKES, RICHARD, of Barkham Square: born, 1746; Ensign, June 10th, 1789; Lieutenant, May 18th, 1790; Captain, November 24th, 1794; resigned, October 20th, 1808. Probably one of the Weekes of Sussex. " Married, 1791, Richard Weeks, Esq., to Mrs. Hill, widow of T. Hill, of Twickenham."

WHEBLE, JAMES, of Bulmershe: eldest son of James Wheble, of Woodley Lodge, by Emma, daughter of Timothy

O'Brien, of Kilcor; Captain, September 15th, 1837; re-
signed, August, 1852; married, 1850, Lady Catherine
Elizabeth St. Lawrence, second daughter of Thomas, third
Earl of Howth; died, 1884.

WHEBLE, WILLIAM FRANCIS, of Bulmershe; younger
son of James Wheble, of Woodley, by Emma, daughter of
Timothy O'Brien; Lieutenant, January 19th, 1853; Captain,
May 23rd, 1853; Paymaster to 97th Regiment, May 18th,
1860; went to the 7th Dragoon Guards, December, 1863;
served in the Egyptian War, 1882—Medal and Khedive's
Star; retired with the rank of Lieut.-Colonel, June, 1885.

WHEBLE, EDMUND, of Bulmershe: Ensign, September
14th, 1837; Captain, December 29th, 1845; resigned, June
13th, 1852.

WHEBLE, TRISTRAM JOSEPH, of Bulmershe: born, 1857;
second son of James Wheble, of Bulmershe, by Lady
Catherine E. St. Lawrence, daughter of the Earl of Howth;
Lieutenant, June 25th, 1877; Captain, April 24th, 1880;
Major, February 2nd, 1889; resigned, December, 1896.

WHICHCOTE, CHRISTOPHER: Governor of Windsor for
Parliament, October, 1642; he was appointed Colonel, 1650;
he delivered up Windsor Castle to the King's forces, December, 1659, when Charles II. was restored to the throne. He
was ancestor of Sir George Whichcote, Bart., of Aswardby,
County Lincoln, who has kindly allowed the portrait of
Colonel Whichcote to appear in these pages. The portrait
is authentic, and Lady Exeter says it was lent to the artist
who painted the frescoes in the House of Lords. This
engraving does not do him justice, for he was an extremely
handsome man, with a rather sad expression in his large
brown eyes. Colonel Whichcote was not a bigot, but
carried out the orders of the Government under whom he
served, and he was right to refuse to allow the funeral of
Charles I. to be made a religious ceremony in opposition
to Parliamentary orders.

WHITE, WILLIAM: Regulating Captain for Berkshire; promoted to Liverpool, 1795.

WHITE, WILLIAM, A. F.: Second Lieutenant, April 2nd, 1890; Lieutenant, March, 1891; Instructor of Musketry, 1896.

WHITEHURST, EDWARD TEMPLE, of Farnborough: Second Lieutenant, March 8th, 1890; drowned at Shrewsbury, July 18th, 1890.

WILDER, JOHN, of Nunhide: only son of Henry Wilder, of Nunhide, by Elizabeth, daughter and co-heiress of Thomas Saunders, of Chaddleworth; Lieutenant, 1758; Captain, 1762; married, Beaufoy, daughter and heiress of Colonel William Boyle; died, 1772, at his son's house at Sulham in his 63rd year.

WILDEGROS [or WILDGOS], READE: Captain, 1625; appointed Muster-Master of Reading, October, 1617.

WILLES, GEORGE COE THOMAS, of Hungerford: born, 1870; eldest son of George Shippen Willes, of Hungerford Park, by Susan Emily, daughter of Thomas Tyrrhitt Drake, of Shardeloes; Second Lieutenant, March 3rd, 1888; Lieutenant, February 7th, 1891; resigned, April 14th, 1891; served for two years with Provisional Battalion, Shorncliffe.

WILLIAMS, CHARLES: born, 1779; Ensign, 1807; Lieutenant, 1807; resigned, 1809. In 1763, a son was born to Charles Williams, of Great Russell Street, Bloomsbury. In 1801, among the list of Government Pensions, appears that of Charles Williams, who received annually £263.

WILLIAMS, CHARLES CROFTS, of Roath Court, Glamorganshire: born, 1866; eldest son of C. H. Williams, of Roath Court, Glamorganshire, by Milicent Frances, daughter of Robert Herring, Esq., of the firm of Williams, Deacon and Co., Bankers; Lieutenant, May 14th, 1885; resigned, February 3rd, 1888.

WILLIAMS, MARTIN: born, 1776; educated at Eton; Ensign, 1803; Lieutenant, 1803; appointed to the 15th

Light Dragoons, October 20th, 1803. Martin Williams, of East Bryngaryan (or Bryngwyn), County Montgomery, married Mary, daughter of John Edward Maddocks, of Vron Iw, and Glanywern.

WILLIAMS, JOHN MAHON, of Reading: born, 1828; eldest son of John Jeffreys Williams, Barrister-at-Law, by Jessie, daughter of Robert Browne, of Jamaica; educated at Eton; Ensign, March 15th, 1853; resigned, January 6th, 1855. His brother was Montagu Williams, Barrister-at-Law.

WINCKWORTH, JAMES: son of Captain Adjutant Winckworth, of Marsh Place, Benham, Newbury; Ensign, June 15th, 1832; resigned, October, 1852.

WOODHOUSE, JAMES: born, 1765; Ensign, 1801; Lieutenant, April 26th, 1803; Captain, 1813; Assistant Surgeon, June 21st, 1801. "Died at Tavistock Place, 1832, by bursting a blood-vessel whilst pulling on a boat, James Woodhouse, aged 56." This may have been some relation as the age differs so widely.

WYKHAM, RICHARD FIENNES: born, 1761; son of Richard Wykham, of Swalcliffe Park, by Vere Alicia, sister of Lord Saye and Sele; Ensign, 1795; Lieutenant, December 27th, 1795; Adjutant, 1803; resigned, July 21st, 1806. Afterwards, apparently, he entered the Church and became Rector of Sulgrave and Chacombe, and married Mary, daughter of Charles Fox.

WYLD, THOMAS (Junior), of Speen: Ensign, September 4th, 1786; Lieutenant, June 15th, 1787; Barrister; died, unmarried, 1789, four months after his father. Speen House was sold in 188- by Captain Wyld, brother of Miss Wyld, of Knotmead, Mortimer.

WYLD, JAMES, of Speen: third son of Thomas Wyld, of Speen; Lieutenant, December 18th, 1793; resigned, 1794; became Rector of Blunsden St. Andrew, County Wilts; married Miss Haverfield; died, 1834. He was probably Regimental Chaplain.

WYVILL, ZERUBBABEL, of Bray: 1762-1837; wrote a March for the Berkshire Militia, 1792, by command of Colonel the Earl of Radnor. He is mentioned as a native of Maidenhead, in David Baptie's *Handbook of Musical Biography*. The last of the family, William Wyvill, Organist of St. Mary's, Maidenhead, died, unmarried, in 1825. In a reply to enquiries in *Notes and Queries*, Mr. William Underhill, of 72, Upper Westbourne Villas, Hove, writes that: "Zerubbabel Wyvill, who composed and published several pieces of music, lived at Inwood House, Hounslow. I saw him there in my boyhood, and duly remember him as an old man, short and thick, with a voice traditionally reported to have been good, but then decidedly the worse for wear. He was twice married; his second wife (who survived him) was Elizabeth, eldest daughter of Thomas Mountford, of Hill End, in the Parish of More, Salop. In 1828, Wyvill was involved in Chancery proceedings, concerning the estate of his father-in-law, by whose will he had been appointed executor. The suit arose out of a family dispute, wherein harmony and the 'concord of sweet sounds,' gave place for a time to 'harsh discords and unpleasing sharps.'"

YEATES [or YATES], ROBERT: born, 1764; Ensign, February 21st, 1799; Lieutenant and Assistant Surgeon, June 19th, 1800; Surgeon, May 25th, 1801; Died, February 11th, 1813.

ZIMENES, SIR MORRIS, of Bear Place; born, 1772; Captain, 1802; resigned, March, 1803; Commanded the Wargrave Rangers Cavalry, 1805, to which he devoted both time and money; High Sheriff of Berkshire, 1805. As a young man he acted with Richard, seventh Earl Barrymore, in his theatre at Wargrave. In 1785, a duel was fought in Hyde Park between Mr. Zimenes and Mr. John Franco. The latter being the challenger, he had first shot, he fired at twelve paces and missed. Mr. Zimenes then fired in the air. At the second shot, Mr. Franco narrowly missed him, again Mr. Zimenes fired in the air, saying he bore no animosity against his adversary. Mr. Franco was asked if he was satisfied, whereupon he replied he "could not fire again at a man who behaved so honourably." The cause of the quarrel never transpired. "Died in Gloucester Place, London, 1822, Lady Zimenes, wife of Sir Morris Zimenes, Bart., of Bear Place, who had married, 1813; she was a widow, Mrs. Cotsford."

Before I close these pages I feel I must express most hearty thanks to all those who have accorded me help in my search for information, especially, Mr. Greenhough, of the Reading Free Library, from whom I have received the greatest kindness and assistance; and I must also mention Miss Dalzell, Mr. Austin, Rev. P. H. Ditchfield, Mr. W Money, Mr. H. Day, Colonel Davis, and Captain Turner, to whom my best thanks are due.

For the information of others engaged in similar research, I add a list of the books from whence information was obtained; and for any description of research work, I can highly recommend Miss Pattie Ostler, 41, Great Russell Street; and as a Typewriter, Miss Sikes, 13, Wolverton Garden, Hammersmith, S.W., both of whom have worked for me in a most satisfactory manner.

## BOOKS SEARCHED TO OBTAIN MATERIALS FOR THIS HISTORY.

Military History; *Clode*.
Calendar of State Papers.
Camden Society's Publications.
History of the Rebellion; *Clarendon*.
Newbury; *W. Money*.
Battles of Newbury; *Money*.
History of Berkshire; *Ashmole*.
Pamphlet on Militia; *Captain Warde*.
Berkshire; *Ashmole*.
History of Wallingford; *J. Kirby Hedges*.
Visitation of Berkshire; *Ashmole*.
History of England; *Collins*.
The Courier.
Reading Mercury, 1723—1896.
Records of the Militia Battalions of the County of Southampton.
History of England; *Hume*.
History of Speen; *W. Money*.
History of the 66th Regiment; *Groves*.
Book of Dignities; *Haydn*.
Berkshire; *Lysons*.
Environs of Newbury.
Pall Mall Magazine, Article, May, 1896; *Lord Raglan*.
Hundred of Bray; *Kerry*.
Reading; *Man*.
Tour Round Reading; *Snare*.
Siege of Donnington Castle.
History of Henley-on-Thames.
Siege of Wallingford.
Extinct Peerage; *Burke*.

Parliamentary Blue Books, Ancient MSS.
Notable Events in Reading; *Guilding.*
Peerage, County Families, &c.; *Burke.*
List of Militia Officers 1779; *Bodleian Library.*
Genealogies; *Berry.*
Parliamentary Papers, House of Lords, and House of Commons,
    1660—1742.
Registers, St. George's, Hanover Square.
Universal Magazine, 1735—1764.
Bygone Berkshire; *P. H. Ditchfield.*
Magistrates and Lieutenancy of Berkshire; *Roberts.*
Gentleman's Magazine.
Reminiscences of an Octogenarian; *Alderman Darter.*
Britannia; *Camden.*
Aldermaston Bowling Club MSS.
Drill Book of the Woodley Volunteers, 1805.
MSS. Collection of Marches in the British Museum.
Northamptonshire and Rutland Militia.
Registers, St. Mary's, Reading.
History of Sherwood Foresters.
Windsor; *Hakewell.*
St. Lawrence's Church, Reading; *Kerry.*
Notable Trials—Family Library.
Calendar of State Papers.
Calendar of Home Office Papers.
History of England; *Smollett.*
The Army and Reserve; *Colonel Adair.*
West Surrey Militia (2nd); *Colonel John Davis.*
The Present State of Great Britain, 1726; *John Chamberlain*
Reading; *Coates.*
Berkshire Archæological Magazine.
Berkshire Chronicles.
Notes and Queries.
History of Swallowfield; *Lady Russell.*
Register of Officers of the Berkshire Militia MSS.
Court-Martial Books of the Berkshire Militia MSS.
History of Berkshire; *Colonel Cooper King.*

Army Lists; *Hart.*
Annual Registers, 1759—1844.
Transactions of the Newbury Field Club.
Quarterly Review.
Life of John Hampden; *Lord Nugent.*
War Office Papers, &c., Record Office.
History of England; *Knight.*
Reading Corporation MSS.
Army Lists, British Museum, and Bodleian Library.
Abingdon Corporation MSS.
Pretenders and their Adherents.
The British Officers; *Stocqueller.*
Marriage Licenses; *Harl. Soc.*
Privy Council Orders, British Museum.
An old Black-Letter History of England; *Edward Hooves.*
Registers, Thatcham Church.
Bedfordshire Militia; *Sir John Burgoyne.*
Royal North Gloucestershire Militia; *W. J. Cripps.*
Marching Order MSS., Record Office.
West Yorkshire Militia; *Captain Raikes.*
Lives of the Admirals of England.
Everyday Book; *Hone.*

# NAME INDEX.

*The Names not numbered will be found in the Officers' List, Chapter XV., which is not indexed.*

Abingdon, Countess, 182.
Abingdon, Fifth Earl,
Abingdon, Sixth Earl,
Abingdon, Seventh Earl,
Acland [or Alland], John Fortescue, 88, 104.
Adam, M. M. Mercer,
Adams, F. E., 229.
Adams, George (Bugler), 172.
Adams, Joseph,
Adams, J. W. R., 208.
Adams, Nicholas, 19.
Agar, Hon. Mrs., 144.
Albans, St., Duke of, Charles B., 62, 63, 75, 78.
Albans, St., Duke of, George,
Alder, W. B., 184.
Aldridge, Captain R., 37.
Alexander, C.
Algeo, John, 142.
Allfrey, F. V., 205.
Allfrey, Mowbray, 194.
Alyn, Thomas, 7.
Allin, John, 194.
Andrews, 67.
Andrews, Captain, 25.
Andrews, James Pettit, 77, 87.
Andrews, Joseph, 76, 87, 155.
Andrews, William, 77.
Angell, W., 65.
Annesley, Francis, 88.
Apthorp, K. P., 205.
Arbuthnot, Robert C., 194.
Archer-Houblon, G. B. E.
Archer-Houblon, H., 229.
Argyle, Duke of, 63.
Arnold, James, 42.
Ashbrooke, Viscount, 125, 144.
Ashurst, C. H., 203.
Assone, John, 21.
Astley, Sir Jacob, 26, 28, 31, 33, 35.
Aston, Sir Arthur, 30, 32.
Aston, Sir Willoughby, 76, 77, 87.

Atkins, Atkins, Edward Martin, 149    *See Martin.*
Atkins, Francis Martin, 194
Atkins, William Hastings Martin,
Atkinson, Charles,
Auger, 129
Austin, John, 174.

Baber, Thomas Draper, 76, 87.
Bacon, Charles, 175, 178.
Bacon, George William, 184.
Badcock, Nicholas, 45.
Bagot, W. H. K., 224.
Bailey, Benjamin 145, 149.
Baker, 65.
Baker, James, 115, 123.
Baker, Lewis, 7.
Banbury, Lieutenant, 119.
Bardsley, James, 139.
Bardye, John, 19.
Barham, T., 38.
Barker, 65.
Barker, Frederick G., 213.
Barker, G. William, 184.
Barker, J., 22.
Barry, Hon. Augustus, 125.
Barry, S. L., 224.
Barrymore, Lord, 71, 124.
Barrymore, Seventh Earl,
Barkstead, Colonel, 38.
Barlow, Frederick Barrington Pratt, 209.
Barlow, Thomas Arthur Pratt, 194.
Barns, T, 65.
Bassett, Francis, 19.
Bates, 25.
Batson, Stanlake, 123.
Baxter, Colonel, 38.
Bayntun, Ed., 29.
Bayntun, Henry, 184.
Beale, Sergt.-Major, 230.
Beales, Benjamin,

Bedwards, Tom B.,
Beke, Thomas, 4.,
Bellas, Joseph Harvey, 104, 115.
Bellasis, Captain, 25, 33.
Berrington, Walter, 12.
Berry, William, 19.
Berkshire, Earl of, 29.
Bertie, Hon. C. C., 194.
Bertie, Hon. M. C. F., 204.
Bever, Samuel John, 174.
Bewell, William, 19.
Birch, F. M., 206.
Birnie, 173.
Birnie, James, 145.
Blackstone, Henry, 123.
Blagrave [or Belgrove], Anthony, 19.
Blagrave, Daniel, 29.
Blagrave, Edward, 178.
Blagrave, Edward (Bugler), 172.
Blagrave, John, 37, 43, 76, 104, 175, 178.
Blagrave, John Charles, 209.
Blagrave, Joseph, 104, 116, 131, 146.
Blagrave, Mrs., 182.
Blagrave, Mrs., 182.
Blagrave, Thomas, 77, 87.
Blake, J. C., 115.
Blandford, Marquis of, 161.
Blandy, Adam, 184.
Blandy-Jenkins,        *(See Jenkins.)*
Blyth, C. V., 194, 204.
Boone, William, 21.
Booth, W., 22.
Borrett, Col., 214.
Boult, John, 76, 77.
Bouverie, Hon. E., 178.
Bouverie, Hon. M.,
Bouverie, Hon. P. L., 153.
Bowles, Colonel, 33.
Bowles, F. R., 184, 188.
Bowles, J. S.
Bowles, T. J., 194.
Bowyer, Sir George.
Boyce, G., 65.
Brackstone, William, 35.
Braham, W. S., 184, 194.
Bray, Colonel, 202.
Breton, General, 184.
Brickman, C. D., 184.
Bristow, Henry,
Brocas, Bernard, 126.
Brocas, Bernard.
Bromley, Hon. H., 145, 149.
Brookland, William, 76.
Brookman, W. L., 149.

Brown, — 225.
Brown, John,
Brown, Tom,
Browne, Andrew, 19.
Browne, Richard, 35.
Bruce, Lord, Charles Bridewell Bruce,
Brummell, William, 129.
Bulley, Edward, 19.
Bulley, F. A.
Bulley, J. B.,
Bunney, Edward Brice,
Burges, Benjamin, 44.
Burne, K. P., 205.
Burne, M. K., 208.
Burnett, B.,
Burningham, John, 45.
Butler, Andrew,
Butler, Joseph, 115.
Butler, Major,
Butler, Thomas William,
Butler, Sergt.-Major, 230.
Byrne, Joseph,

Cameron, General, 205.
Camyll, Roger, 21.
Cane, Robert, 142, 149.
Cannon, Captain,
Cardiff, William,
Carey,
Carlingford, Lord,
Carpenter, General, 60.
Carter, 22.
Cazenove, P., 226,
Cazenove, R. F., 215.
Cerjat, A. S. De,
Chamberleyne, A.,
Chapman, John,
Chauval, Edward,
Clanchy, Richard,
Clarke, John,
Classon, Henry,
Claver, Joseph, 44,
Claveland [or Cleveland], William, 115.
Clerke, Sir E., 24.
Clifford, Thomas, 22.
Climenson, H. J. M.
Coanes, 22.
Coate, W., 65.
Cobham, Alexander C., 176.
Codd, R. B.,
Coker, William, 7.
Coleman, G. T., 178.
Coles, William, 174, 178.
Collett, William, 19, 21.
Collis, William, 150.

Z

Collyer, John,
Compton, Sir W., 65.
Cooper, Edward, 21.
Cooper, Robert, 22.
Cope, William, 52.
Cordery, John, 7, 19.
Costobadie, G. E., 206.
Cowell, C., 65.
Cox, Francis Renell, 182.
Craven, A. W., 201.
Craven, Earl of, 99.
Craven, Earl of,
Craven, Hon. R. C.,
Craven, Hon. William,
Craven, O. W.,
Cray, Drummer, 172.
Creed, Thomas, 21.
Croft, H. H.,
Croft, W., 155.
Croome, Isaac, 21.
Crowe, David,
Curtis, Thomas John,
Curtis, Captain, 36, 39.
Cutler, Moses, 7.

Daling, Edmund, 21.
Dalmer, Francis,
Dalzell, Robert,
Damant, Guybon,
Danvill [or Darvill], Charles,
Dauncey, — 229.
Davenport, J. T., 174.
Davies, George,
Davies, E., 203.
Day, 43.
Deane, Arthur, 178.
Deane, H., 65.
Deane, H. B., 125.
Deane, John, 77.
Dentry, E., 65.
Desborough, John,
De Vitre, H. D.,
Dickson, Colonel, 211.
Dixwell, J. 49.
Dodd, 145, 149.
Dodd, John, 76, 79, 87, 91.
Dodwell, J. W.,
Doe, 45,
Dole, John, 7.
Doleman, John, 19.
Dolman, Thomas, 52.
Donovan, Capt., 221.
Doran, James Goddard, 174.
Douglas, James, 184.
Downes, Percy, 215.

Draper, Sir T.,
Draycott, 65.
Drummond,
Duffield, C. J E.
Dunn, Sergt.-Major, 230.

East, A. H., 145, 153.
East, G. F. Clayton, 201.
Edmundes, Justman, 19.
Edwards, C. M.
Edwards, T. H., 150.
Edwards, Thomas Hughes,
Effingham, Earl of
Egerton, A. G.
Elkins, W., 65.
Elliott, G. H.,
Elliott, Thomas,
Elwes, George, 104.
Erneley, 52.
Essex, Earl of, 30, 32.
Est, Richard, 7.
Evans, — 229.
Evans, Henry, 88, 104, 115.
Evelyn, Arthur, 28, 43. 45, 41, 49.
Everett, F.,
Evett, Thomas, 7.
Eyre, G. B., 194.  *(See Archer Houblon).*
Eyre, Matthew, 40.
Eyston, Francis T., 194.

Fairfax, Sir T., 37.
Fennel, Edward,
Feversham, Lord, 57, 59.
Fielding, Colonel, 32.
Fincher, General, 40.
Finucane, Michael,
Fitzroy, Lord C., 159.
Fleetwood, General, 46.
Floyer, Peter, 76, 87.
Folkestone, Viscount, 174.
Fonblanque, John, 115.
Forrest, Arthur,
Forrest, Sandford,
Forrest, Thomas,
Fortesque, Richard, 36.
Fowler, Butler, General, 217.
Fowler, Ernest M.,
Frazer, General, 113.
Freman, Henry, 7.
French, W. N., 104, 115.
Frost, Walter, 49.

Garnett, Taylor,
Garraway, Eleanor, 65.

Gateley, John, 7.
Gayger, R., 22.
Gent, Nicholas, 7.
Gibbs, John, 99.
Gifford, Captain, 21.
Gilbert, Mary,
Gilbert, Richard, 12.
Giles, J., 65.
Gill, 145.
Gill, James, 115.
Gill, Philip, 104, 115.
Goddard, Richard, 45.
Goddard, Vincent, 36, 39, 43.
Goffe, Colonel,
Golden, R., 65.
Goodlake, Thomas, 155.
Gower, C. C. Leveson,
Gower, John Leveson, 171.
Gower, Philip Leveson, 178,
Graham, G. B.,
Gray, J. R., 211.
Green, Captain, 112.
Green, Captain,
Greene, Richard, 43.
Greenhead, Charles, 175, 178.
Greenway, Henry, 178.
Grenfell, G. G.,
Grenfell, W. H., 194.
Gribble, John,
Griffith, Christopher W. Darby, 204.
Griffin, 65.
Griffin, John,
Groves [or Graves], Thomas, 88, 104, 115.
Grubb, George,
Guy, George, 149, 175.
Guyenett, 145, 149.

Hackett, Patrick, 19.
Hallett, G. H.,
Hallett, William,
Hamilton, Arthur,
Hamilton, Colonel, 24
Hammond, Captain R., 24, 33, 39, 44.
Hampden, John, 30, 32, 33.
Hance, James, 174, 178.
Handasyde, Colonel, 65.
Hankey, S. A., 194.
Hankey, S. T.,
Hannington, Richard, 65.
Hargreaves, Arthur, 203.
Hargreaves, Robert, 215.
Harrington, J., 49.
Harrison, Colonel, 40, 43.
Harrison, J. S.,

Harrison, Thomas, 36, 38.
Harvey [or Hervey], Lionel Charles, 174.
Harvey, Peter, 22.
Hatch, George, 76.
Hatt, John, 19.
Hawes, Francis 115, 133.
Hawker, General, 164.
Hawkins, Benjamin, 174.
Hay, A. W. H., 206.
Hay, H. C. F., 218.
Hayes, John B.,
Heath, James,
Heddige, John, 72.
Henderson, H. G.,
Hensman, Richard, 7.
Herbert, Colonel, 209.
Herbert, General, 196, 198.
Herbert, Lord, 52.
Hercey, T. F. J. L.,
Hercey, T. H. G.,
Hertford, Marquis of, 29.
Hill, 145.
Hill, John,
Hill, Swann, 126.
Hippesley, W. H., 194.
Hodge, Baldwin,
Hodgson, Thomas, 121.
Hodgson, W. S.,
Hodgson, W. S.,
Holdsworth, F. R., 149, 151.
Holden, H. W., 203, 215.
Holland, Henry Richard, Earl of, 26, 27, 30, 33, 41.
Holland, S. E.,
Holland, Thomas,
Holloway, 65.
Holloway, Benjamin, 145.
Holmes, General, 80.
Holt, George,
Hollyer, 225.
Homan, Henry, 21.
Homfray, H. R., 211.
Homfray, J. G. R., 206, 209.
Hood, John, 19.
Hopkins, E. G. R., 198.
Hopton, John, 7.
Hopton, Lord, 31, 33.
Horne, Arthur, 43.
Horwood, J. J.,
Hoskins, Matthew, 7.
Houblon. *(See Archer)*.
Houghton, T. A.,
Howard, Colonel, 70.
Howard, Thomas,

Hudson, Edward, 19.
Hunter, Sir C. S. P., 183.
Hunter, C. R., 201.

Imhoff, Charles, 126, 129, 149.
Ince, H. R., 125.
Ince, Henry, 178.
Ince, James, 174.
Isherwood, Richard,
Ivery, William, 37.

Jenkins, John Blandy, 194.
Jenkins, John Blandy,
Jenkinson, Rev., 65.
Jekyll, John,
Jerome, John, 19.
Johnson, Thomas, 19.
Johnston, A., 49.
Johnstone, C.,
Johnstone, Lieutenant-General, 105.
Jones, John,
Jordan, Colonel, 204.
Judd, Stephen, 174.
Justice, Thomas, 76, 87.

Kates, 172.
Kearney, H. J., 161.
Keepe, Andrew, 43.
Kenton, John, 22.
Kenrick, William,
Kenrick, 54.
Keppel, General, 102.
Key, Thomas, 150.
Kinge, John,
King, A. H. W.,
Kinnersley, W. T.,
Kirke, Lewis, 32.
Knolles, Sir Francis, 20, 21, 27.
Knollys, Captain,
Knox, A.,
Knipe, Christopher,

Lambe, W., 103.
Lambert, General, 45.
Lamport, 65.
Lane, J. H., 184.
Lang, F. H., 184, 195.
Langford, Joseph, 88.
Lee, H. P., 125, 129.
Lee, G. P., 178.
Lee, Robert, 65.
Lennox, Lord G., 100.
Lenthall, 129.
Leverington, Captain, 36.
Leveson-Gower, 178.

Lewis, C., 204, 221, 224.
Leycester, H. H.
Leycester, O. W.
Lidyard, John, 21.
Lindsey, Earl of, 28.
Linscome, Mrs., 65.
Lintall, 129.
Littlepage, William, 21.
Lloyd, William,
Loveden, Edward Loveden, 104, 128, 129.
Loveden, Pryse, 129, 144.
Lovelace, Baron, 53, 54.
Lovelace, Sir Richard, 20.
Lower, Lieut.-Colonel, 35.
Loyd-Edwards, Mrs.,
Ludd, Ned., 164.
Lush, John, 45.
Lynch,
Lyppescombe, William, 12.

Macpherson, Lachlan, 149.
Mackworth, Praed, 91.
Madocks, John Edward, 125.
Maitland, Thomas Fuller, 184.
Mansfield, Count, 18.
Marsh, Sir C., 131.
Marsh, W., 129.
Martin-Atkins, Atkins Edward,
Martin-Atkins, William Hastings,
Martin-Atkins, Francis,
Martin, H., 30.
Maton, J.,
Mauncell, Colonel, 202.
Maurice, D. B., 211.
Max, Zacharias, 19.
May, 65.
Mayes, William,
Mayle, William, 7.
Maynard, James, 36.
Mears, Ellis,
Medley, A. L.
Meter, J. C., 126.
Meyrick, Samuel,
Micklin, Robert,
Milne, John, 184, 188, 195.
Milman, W. G.,
Moody, 65.
Monck, W. B.,
Monckton, General, 109.
Monmouth, Duke of, 57.
Montagu, Ed., 82.
Morland, W. H., 194.
Morland, G. W., 184.
Moreton, John,

Morres, E. J.,
Morrice, Charles, 123.
Morris, Capt., 37.
Morrison, 195.
Morshead, W. E.,
Mowbray, R. G. C., 194.
Moyle, John, 19.
Mullyns, John, 19.
Murphy, T., 200.

Nepean, C. E. M. Y.,
Newbolt, F. N.
Newbolt, J. T., 123.
Newbolt, William,
Newbury, John,
Newton, Edward,
Neville, Col Richard. 34.
Neville, Richard Aldworth, 104.
Neville, Hon. Richard,
Nicholls, John, 21.
Norreys, Lord. 182, 206. *See* Abingdon.
Norris, H. C.,
Norris, John, 174, 178.
Norris, Nicholas, 7.
Noyes, Thomas Buckeridge, 76.

O'Brien,
Oldfield, C. C., 194.
Ormond, Duke of, 62, 63.
Osgood, Lawrence Head, 76, 87.
Osborn. William,
Otway, C. C.,
Otway, J. T. F., 203, 208.
Owens, Sergt., 203.

Page, F., 161.
Paget, Lord, 125.
Parker, John, 174, 178.
Parker, Robert, 104, 115.
Payn, James, 121.
Payne, Luke, 21.
Peacock, — 52.
Pearson, C. L. M., 201.
Pearse, William, 21.
Pechell, E. R. C.,
Pechell, W. M. C., 194.
Penny, Drummer, 172.
Pepper, Major-General, 65.
Perse, John, 19.
Percv, Lord, A. M. A., 207., 209.
Phillips, G. E.,
Phillips, John,
Phillips, William, 174.
Pinkney, George,

Pitt, General, 117, 118.
Pittman, Sergeant, 99.
Pocock, D., 65.
Pocock, J., 65.
Pococke, John,
Pococke, John Blagrave, 145, 174.
Pole, Henry, 178.
Poleman, William, 21.
Poole, Richard, 21.
Popham, F. W. Leyborne, 208.
Porter, Andrew, 21.
Porter. Frederick, 211.
Powell, Arthur Annersley, 123.
Powell, Thomas, 21, 22.
Powell, Roger, 22.
Powney, Portlock Pennyston, 104, 115, 129.
Powys, Philip, 100.
Praed, William Mackworth, 76, 87, 91.
Pratt, Lord George M.,
Prestage, Thomas,
Preston, Colonel, 66.
Preston, John, 210, 215.
Preston, Sergeant-Major, 171.
Price, J. C., 104, 115, 119.
Prower, Major, 186.
Pryer, Christopher, 21.
Pulleyn, William, 14.
Puntor, T., 65.
Purcell, John,
Purnell, E. K.,
Purvis, Edward, 175.
Pye, Henry James, 104.
Pye, Sir Robert, 29.
Pye, Walter, 104. 115, 126, 153.

Radnor, Earl of. 138, 147.
Ramsay, Colonel, 22, 23.
Ramsey (Ensign), 145.
Ravenshaw, T. W., 126. 152, 166, 170, 174.
Ray, 65.
Reed, H. W., 184. 195.
Reed, Superintendent, 185.
Reeves, Edwin,
Reeves, John, 76, 87.
Reid, G. A. C., 104.
Reille, Elizabeth, 103.
Richard, J. 22.
Rickman, S.,
Rhodes, 65.
Rhodes, H. V.,
Rhodes, J. E.,
Robbins, John,
Robinson, General. 159.

Roe, William, 174, 178.
Rokeby Baron.
Rooke, General, 143.
Round, Henry,
Rouse, Richard,
Rows, William, 14.
Royds, A. H.,
Rudland, Jones,
Rupert, Prince, 30, 33.

Saunders, 54.
Saunderson, Anthony,
Savernake, Viscount, 208.
Sawyer, Charles,
Sawyer, John,
Saxton. *(See Sexton).*
Schrader, F. H., 149.
Scott, H.,
Seagrove, John,
Seely, Sergt.-Major, 230.
Sellwood, Richard, 76, 87.
Sexton, Clement, 87, 104, 115.
Seymour, 182.
Seymour, E,. 76, 87, 104.
Seymour, R.,
Shackell, William,
Shackell, William R., 184.
Shaftesbury, E., 7, 82.
Shaw,
Shaw, Lieutenant-Colonel,
Sheppard, Edward, 104, 115.
Sherren, William, 149
Sherson, Alexander Norvell, 175, 177.
Sherwood, Hugh, 19.
Sidney, Philip,
Simpson, Francis, 123.
Sladden, W., 104, 115.
Slade, W.
Slocock, C. S., 183.
Slocock, Edmund, 174.
Smith Bandmaster, 172.
Smith, Benjamin,
Smith, J. W., 184.
Snook, Thomas, 150.
Southby, Samuel, 87.
Southey, T., 65.
Speed, D. W. H.,
Spurgeon, General, 209.
Sheen, James,
Squires, William,
Staden, Sergeant-Major, 188. 229.
Stapper, Robert, 4.
Stares, William, 182.
Stark, Charles George, 104.
Statham, Roger, 14.

Stead, Major, 146.
Stephens, Henry,
Stephens, M. H. G., 125, 129, 174.
Stephenson, John, 104.
Stephenson, Rowland,
Stephenson, William, 43.
Stewart, F. T., 211.
Stonor, Cornet,
Stracey, Thomas, 129.
Stratton, W., 174.
Stuart, John,
Sturges, G. T.,
Sturges, J. H., 174.
Sturges, John,
Styles, Clement, 88.
Sudley, Viscount, 214.
Suffolk, Duke of, 5.
Sykes, Sir Francis, 138, 142, 149.

Taff, John, 7.
Taylour, Thomas, 19.
Tebbott, Robert, 183.
Temple, C. P., 207, 211.
Thatcham, 65.
Thackwell, Major-General, 199.
Theobald, F. C., 210.
Thomas, Stephen,
Thomson, E. T., 183.
Thornton, W.,
Thorowgood, Sir John, 43.
Thoyts, M. G., 176.
Thoyts, William,
Thoyts, W. R. M., 184, 185, 190, 226.
Throckmorton, P. H.
Timbrell, W. H., 116.
Toft, John, 14.
Toll, A. N., 115.
Toogood, C. F. S. G., 205.
Toovey, — 65.
Towsey, William, 76, 88.
Tracy, Lord, 52.
Treacher, George, 149.
Trevor, Henry,
Tristram, L. S. B., 203
Tubbe, Roger, 19.
Turner, Charles, 225.
Turner, J., 65.
Tylden, Osborne, 115.
Tylby, Richard, 12.

Uffington, Viscount,
Urlwin, J.,

Vachell, — 30.
Vachell, Sir Thomas, 21.

Van de Weyer, B. G.,
Van de Weyer, V. M. B., 194, 206.
Van de Weyer, W. J. B., 215.
Vane, George, 52.
Vane, H., 49.
Vanlore, Sir P., 37.
Vansittart, Arthur, 76, 100, 104, 115, 145.
Vansittart, Arthur, 87, 95.
Vansittart, G. H., 149, 153.
Vansittart, H.,
Vansittart, Henry,
Vansittart, N. H., 204.
Vansittart, Laura, 153.
Velley, Charles,
Velley, Thomas, 104.
Vere, Lord,
Vincent, H. W.,
Vincent, C. E. H., 194.
Vincent, E., 204.
Voules, C. S., 184.

Wadling, W. A., 226.
Wallingford, Lord, 20.
Wallis, John, 104, 115.
Wallop, R., 49.
Walker, C. H., 150.
Walker, John,
Walter, John, 76, 87, 104. 115.
Walter, John Abel,
Walter, J. B.,
Wantage, Baron, 231.
Warneford, Francis
Warneford, — 65.
Wayland, J. S.,
Watson, — 119.
Webb, John, 36.
Webb, William, 19.
Webster, J., 22.
Weekes, Richard,
Weyland, J. S.,
Weymouth, Lord, 96.

Wheble, Edmund,
Wheble, James, 176, 178.
Wheble, T. J., 203.
Wheble, W. F. J., 184.
Wheeler, Trevor, 131.
Whichcote, Christopher, 43, 45.
Whitaker, — 179.
White, John, 14.
White, William,
White, W. A. F., 215, 225.
Whitehurst, E. T., 215.
Whitelock, General, 152.
Wigmore, Thomas, 19.
Wilder, John, 76, 87.
Wildgros, Reade, 20.
Willes, Capt., 182.
Willes, G. O. T., 218.
Williams, C. F., 142.
Williams, Charles,
Williams, C. C., 211.
Williams, Martin,
Williams, J. M.
Willis, General, 66.
Winckworth, James, 178.
Withwall, John, 14.
Wirge, Thomas, 19.
Wise, Dr., 172.
Woodhouse, James, 174.
Woodare, John, 19.
Wykham, Richard James,
Williamson, George, 19.
Wyld, Thomas,
Wynn, Sir Watkin, 71.
Wyld, James, 123, 129.
Wylmore, George, 19.
Wyvill, Z., 127.

Yates, Rev. S. W. 182.
Yeates [or Yates], Robert, 150.

Zimenes, Sir M.

# INDEX OF PLACES.

*This Index is not a complete list of all the Places ; those not numbered will be found in the Officers' List.*

Abingdon, 25, 29, 31, 39, 65, 85, 201.
Aboyne,
Adarley Common, 107.
Acton, 102.
Adderbury, 109.
Aldershot, 189.
Alresford, 151.
Aldermaston, 157.
Andover, 82.
Appleford,
Arborfield, 140,
Arundel, 131,
Ascot,
Ashford, 157, 221.
Ashdown, 125, 216.
Athlone, 168.
Aylesbury, 46.

Banbury, 106, 109.
Barkham, 137.
Barnet, 111.
Basing, 35.
Basingstoke, 90, 108.
Basildon,
Bashford, 163.
Bath, 145.
Bear Place,
Bearwood,
Beaurepaire,
Beenham, 21.
Benham, 21, 99.
Berryhead, 240.
Beverley, 28.
Bexhill, 168.
Bicester, 110.
Bideford, 167.
Billingbear, 43.
Bill Hill,
Bilston, 173.
Bincombe, 154.
Binfield, 65.
Bisham,
Bishopsgate, 17.

Blackburn, 167.
Blackheath, 55, 102.
Blatchingdon, 123, 133.
Bletchley,
Blewbury, 137.
Bloxham, 109.
Bow, 111, 112.
Bracknell, 126.
Brackley, 26.
Brasted, 113.
Bray, 127.
Brenchley, 114.
Brentford, 31.
Bridgewater,
Brighton, 126, 131, 137.
Bristol, 33, 63, 146.
Brixham, 139.
Bromley, 103, 112.
Brompton, 117.
Buckland,
Bulwell, 162.
Bulmershe, 100.
Burford, 111.
Buryhead, 139.
Buxton, 166.
Buscot, 129, 137.
Bushey,

Calcot,
Caversham, 145.
Cawsand, 139.
Chalgrove, 33.
Chapel House, 110.
Charridge, 77.
Charlbury, 137.
Charlton,
Chatham, 117.
Chelmsford,
Chieveley,
Chipping Norton, 108, 110.
Cholsey, 12.
Churn, 217, 244.
Clerkenwell,

Clifton, 164.
Clifton Moor, 71.
Cookham,
Coxheath, 101, 102, 117.
Colchester, 69.
Coleshill,
Colwart,
Colne,
Culnbrook, 31, 118.
Corfe,
Corfu, 184.
Cork, 168.
Cowes, 53, 54.
Cranbrook, 114, 116.
Crookham,
Crutchfield,
Cuckhamsley, or Cutchinsloe, 23.
Culham,
Culloden, 71.

Darsham,
Dartmouth, 141.
Daventry, 26, 28.
Deal, 6.
Deanwood,
Denford,
Derby House, 40.
Derby, 71, 166.
Derbyshire, 163.
Devizes, 80.
Doddington,
Domingo House, 242.
Donnington, 35, 36, 251.
Dover, 124, 136.
Dublin, 146.
Dunstable, 46.

Eastbourne, 129.
Earley, 65.
Edgware, 111.
Egham, 103.
Ellcott, 254.
Englefield, 216, 218.
Erleigh,
Eynsham,

Farley Hill
Faringdon, 108.
Fawley,
Fecamp,
Finchley, 71.
Fishcombe,
Fishguard, 140.
Folkestone,
Forbury, 18, 161.

Forest, 25.
Fort Barracks,
Fox Hills, 202.

Galway, 168.
Gillingham, 117.
Gloucester, 52.
Gloucester Lodge, 147.
Goodrest, 182.
Goudhurst, 113, 116.
Greywell,
Greenlands,
Greenwich, 62, 102.
Guildford, 40.

Hadlow, 114.
Hagbourne, 145.
Hailingbury,
Ham, 119.
Hammersmith, 107.
Hampstead Marshall,
Hampstead Norris,
Hampstead, 111.
Hampton Court, 32.
Hanworth, 250.
Hannay, 152.
Harewood Lodge,
Hardwick, 100.
Hare Street, 107.
Hastings, 123.
Hawkhurst,
Heathfield Park, 132.
Henley, 102, 107.
Hendred,
Hennerton,
Hessian Camp,
Highgate, 111.
Hilsea, 108.
Hornsey,
Horsemunden, 114.
Horsham, 167.
Hounslow, 56, 57, 58, 69, 107.
Huddersfield, 166.
Hull, 28.
Hungerford,
Hurley, 71, 145.
Hurst,
Hythe, 133.

Ightham, 112.
Ilford,
Ilsley, 82.
Inner Temple,
Ipswich, 159.
Ireland, 146, 148.

Isle of Wight, 24, 53, 69.
Islip, 110.
Ives Place,

Kensington, 107.
Kidlington, 110.
Kingdown,
Kingston Bagpuize,
Kingston Lisle,
Kingston-on-Thames, 30, 31.
Kirklington, 110.
Kitts End,

Lamberhurst,
Lambeth, 112.
Lambourne, 77.
Lancashire, 166.
Leeds, 166.
Lenham, 113, 116.
Lenton, 164.
Letcombe Basset,
Lewes,
Limerick. 169.
Liverpool, 170, 172.
Lockinge,
London, 28, 49.
Londonderry, 169.
Longford Castle, 138.
Luckley House, 158.
Lydd, 220.
Lynne, 57.
Lymington, 152.

Macclesfield, 166.
Maiden Erleigh,
Maidenhead, 173.
Maidenhatch,
Maidstone, 112, 116.
Mallings, 103.
Manchester, 166.
Marlborough, 80.
Marlow, 129.
Marcham,
Margate, 137.
Merchant Place, 179.
Middleton, 168.
Milford Haven, 14.
Milk House Street, 114.
Mill Bay, 167.
Milton,
Milstead,
Mitchenhampton, 199.
Mortimer,

Netley, 151.
New Lodge,

Newbury, 19, 22, 33, 47, 65, 83, 95,
    99, 118, 120, 121, 144.
Newry. 169.
New Sarum, 126.
Newthorpe, 109.
Newton, County Cambridge.
Norman Cross,
Northamptonshire,
Norbiton Hall, 128.
Nore, 148.
Northall, 110.
Nottingham, 28, 99, 161.
Nunhide, 77.
Nutley, 85.

Oakfield,
Oakingham, 65, 90.
Offham, 102.
Overton, 83, 86.
Oxford, 28, 32, 34, 63, 65.

Paddington. 112.
Padworth,
Park Place,
Peasemore, 77.
Pigeon House, 146.
Plymouth, 20, 53, 101, 167.
Poole, 145.
Portsmouth, 53, 101, 109, 149, 150,
    184, 186.
Portsea,
Potter's Bar, 110.
Preshute, 80, 85.
Preston, 66, 167.
Purley, 133.
Putney,

Radford, 163.
Radley, 25.
Raglan, 52.
Ramsgate. 137.
Reading. 65, 70, 85, 103, 105. 118, 121,
    122, 126, 156, 160, 170, 171. 203.
Redan Hill, 205.
Ridge Mins,
Riverhead, 103, 112.
Rochester, 118.
Romney, 219.
Romsey, 126.
Rumford, 107.
Ruscombe, 77,
Ruddington, 164.
Rushmere, 159.
Rushmoor, 202.
Rye, 124, 220.

St. Leonards, 255.
St. Pancras, 111.
St. Lawrence's Church, 171.
St. Quentin's, 12.
Salisbury, 58, 138.
Salisbury Plain, 195.
Sandgate, 136.
Sandhurst, 65.
Scotland, 7.
Seaford, 125.
Seal, 103, 112.
Sevenoaks, 113, 116.
Sedgemoor, 57.
Shaftesbury, 145.
Shaw, 62, 94.
Shaw Mill, 94.
Sheffield, 166.
Shorncliffe, 136.
Shiplake,
Shottesbrooke, 77, 145
Shoreham, 125.
Shrewsbury, 136.
Shrivenham,
Sindlesham, 137.
Slinfold,
Slough, 105.
Sommerton, 167.
Sonning, 65, 256.
Southcote, 77, 199.
Southampton, 115, 126, 131, 136, 151.
Sparsholt,
Speen, 118.
Speenhamland, 118.
Stanford Place,
Stanlake, 254.
Stanmore, 111.
Steyning,
Stoke Barracks,
Stratford Avon, 108.
Stroud, 102, 117, 198.
Sulhampstead, 227.
Sundridge,
Sunninghill, 120, 254.
Sunningdale,
Suttons, 85.
Sutton Courtney, 77.
Swalcliffe,
Swallowfield, 77.
Taunton, 160.
Temple, 126.
Theale, 33, 146, 216.
Thatcham,
Thanet, 137, 138.
Tidmarsh, 37.
Tilbury, 14, 53.

Tilehurst, 19.
Torbay, 58.
Totterdown,
Totness, 141.
Tonbridge, 116.
Tunbridge Wells, 114.
Trotton, 130.
Tuam, 169.
Vauxhall, 112.
Vron Iw,
Vido, 186,
Wadley, 77, 253.
Wadhurst, 114.
Wallingford, 34, 46, 139.
Wantage, 77, 119, 120, 139.
Wargrave, 255.
Warfield, 65.
Warley, 101.
Warrenpoint, 169.
Watchfield,
Waterdown, 123, 125.
Waterloo, 170.
Weeley,
Welford, 252.
Wellhouse,
Westerham, 113, 116.
Weymouth, 145, 147, 152.
Whetstone,
Whitley, 21.
Whitley Wood, 79.
White Place,
White Waltham, 125.
Whitchurch,
Wilts, 52, 54.
Winchester, 21, 86, 90, 100, 108, 126, 150, 153.
Winkfield, 256.
Wingfield, 65.
Windsor, 31, 43, 45, 102.
Witney, 83, 109, 102, 111.
Wokefield,
Wokingham, 65, 102, 139.
Woodstock, 106.
Woolhampton,
Woolmer Forest, 195.
Woolwich, 126.
Worcester, 46, 52.
Wrotham, 116.
Wytham,
Yarmouth,
Yarmouth, I. W., 54.
York, 166.
Yorkshire, 164.
Zante, 185.
Zealand, 18.

# INDEX OF REGIMENTS.

Abingdon Volunteers, 195.
Albemarle's Dragoons, 99.
Army Service Corps, 221.
Artillery, 185, 283.
Baxter's Regiment, 38.
Bedfordshire Militia, 82, 210, 306.
Berkshire, 119.
Berkshire Yeomanry, 213.
Berkshire Volunteers, 288.
Beynhurst Volunteer Brigade, 291.
Blues, 181, 214, 282, 318.
Bowstreet Officers, 164, 173.
Brunswick Fencibles, 277.
Buckinghamshire, 82, 231.
Buffs, 185.
Bunny Yeomanry, 165.
Cambridgeshire, 169.
Canadian Rifles, 304.
Carbineers, 268.
Carmarthen Militia, 101, 169.
Carnarvon Rifles, 204, 253.
Cheshire Militia, 145, 148.
Coldstream Guards, 204.
Cornish Miners, 151.
Craven's, Lord, Foot, 125.
Cumberland Militia, 170.
Devonshire Militia, 129, 133, 147, 154.
Dorset Militia,
Dorsetshire Regiment, 256.
Dragoons, 60, 80, 101.
Dragoons—1st, 101.
—— 2nd, 101.
—— 3rd, 101, 147.
—— 5th,
—— 6th, 101, 268.
—— 11th and 22nd, 108, 129.
—— 15th,
—— 16th, 312.
—— 20th, 276.
—— 21st, 159.
—— 25th, 270.
Dragoon Guards—5th, 208, 303.
—— —— 7th, 101, 156, 159.
Essex Militia, 136.
Engineers, 302.
Fawley Light Dragoons, 151.

Foot Guards, 80, 101.
Garrison Battery—4th, 289.
—— —— 7th, 283.
—— —— 9th, 271.
Glamorgan Militia, 101.
Gloucester Militia, 86, 147, 199.
Grenadier Guards, 80, 206, 208, 210,
 281, 304, 306.
Greys, 154, 166.
Guards, 80, 253, 276, 277.
Guernsey Militia, 289.
Hants Militia, 46, 86, 147.
Handasyde's, 65.
Hanover and Hessian, 74.
Hereford Militia, 159.
Highland Light Infantry, 213, 317.
Horse Grenadier Guards, 80.
Horse Artillery, 159.
Howards, 70.
Hungerford Cavalry, 182.
Hussars—7th, 181.
—— 10th, 224, 253, 265.
—— 15th, 198, 250.
Independent Co's, 80.
Irish, 80.
Isle of Wight, 53.
Indian Civil Service, 267.
Kent Militia, 69, 187, 216, 219.
Kent Yeomanry, 306.
Kerry Militia, 270.
Lancers—12th, 292.
Lancashire Militia, 133, 136, 187.
Lancashire, 270.
Lennox, Lord G., 100.
Lincoln Militia, 169.
Light Infantry—13th, 230, 295.
Life Guards, 72, 209, 215, 263, 274,
 283, 285, 289.
London Rifle Rangers, 195.
Marines, 256.
Middlesex, 54, 159, 185, 187, 215, 249.
Military Train, 304.
Monaghan, 169.
Native Infantry—31st, 300.
New Romney Fencibles, 133.
Norfolk, 250.

Northampton, 187.
Nottingham, 99, 162.
Northumberland Fusiliers, 210, 304.
Oxford Militia, 123, 126, 133, 185, 187, 199, 213, 231.
Pembroke Militia, 101.
Prince of Wales', 137.
Queen's Bays, 162.
Queen's Westminster,
Reading Volunteers, 162.
Regiment—1st, 187.
——— 2nd, 101.
——— 4th, 268, 269, 286, 289, 307, 311, 316.
——— 6th, 101, 298, 307.
——— 9th, 269.
——— 10th, 272, 297, 311.
——— 11th, 263.
——— 13th, 230.
——— 14th, 101, 253, 260, 274, 317.
——— 15th, 151, 269, 286.
——— 17th,
——— 18th, 101, 207, 252.
——— 20th, 264.
——— 23rd, 195, 272.
——— 25th, 101, 205, 253, 276.
——— 26th, 264.
——— 27th, 270.
——— 30th,
——— 33rd, 280.
——— 34th, 293.
——— 35th, 204, 223.
——— 36th, 291.
——— 43rd, 313.
——— 44th, 311.
——— 46th, 230.
——— 47th, 194.
——— 49th, 202, 206, 208, 303.
——— 52nd, 205, 276.
——— 53rd,
——— 56th, 204, 268.
——— 58th, 253.
——— 59th, 101.
——— 63rd, 268, 311.
——— 65th, 101.
——— 66th, 202, 206, 301.
——— 68th, 185.
Regiment—69th 101, 299.
——— 71st, 317.
——— 72nd, 177, 204, 207, 267, 298, 314.
——— 78th, } 196, 198, 251, 309.
——— 79th, }
——— 81st, 206, 272.
——— 85th, 302.
——— 93rd, 196.
——— 94th, 183.
Rifle Corps, 154, 214.
Rifle Brigade, 205, 218, 288, 304, 310.
Rifles, 60th, 310.
Rifles, Irish, 212, 281, 287.
Rifle Brigade—7th, 229.
Royals, 101.
Royal American, 135.
Scots Greys, 166, 195, 286.
Scotch Fusiliers, 179, 193.
Sherwood Foresters, 162, 216, 280.
Shropshire Militia, 147, 159.
Somerset Militia, 101.
Stafford, 187.
Stafford Militia, 108, 154.
Suffolk, 159.
Sussex, 46, 210, 216, 219, 301.
Surrey, 46, 157, 208, 214, 216, 219, 221, 266, 285.
Tower Hamlets, 199.
Tyrone Fusiliers, 250.
Wantage Volunteers, 152.
Warwickshire Fencibles, 133.
Warwickshire, 212, 280.
Warwickshire Militia, 136.
War Train, 301.
Welsh Fusiliers, 284.
West Kent Yeomanry, 210.
Weymouth Volunteers, 147.
Wiltshire Militia, 185, 186, 187, 199, 267.
Woodley Volunteers, 273.
Worcester, 229.
Wyke Independent Fusiliers, 147.
Yorkshire, 88, 187, 320.
York Hussars, 154.
York Rifles, 166.
York Rangers, 316.

Lightning Source UK Ltd.
Milton Keynes UK
UKHW022356090223
416721UK00001B/199